VIROLOGY

MJP PUBLISHERS

VIROLOGY

P. SARAVANAN
Senior Lecturer
Department of Microbiology
VEL's College of Science
Chennai

Chennai   Trichy   Tirunelveli   NewDelhi

MJP Publishers

All rights reserved
Printed and bound in India

No. 44, Nallathambi Street,
Triplicane, Chennai 600 005

MJP 013   © Publishers, 2020

Publisher : C. Janarthanan

FOREWORD

It is indeed with a great sense of pleasure and privilege that I give this foreword to the book *Virology* written by Dr. P. Saravanan, a dedicated teacher in Microbiology and Biotechnology. Microbiology, at present, has become an exceptionally broad discipline fascinating areas of biological knowledge. Advances in this field are immense and have made this science quite interesting.

There was a long felt-need to have a comprehensive textbook on Virology highlighting the important aspects of basic and systematic virology for the students pursuing undergraduate and postgraduate courses in Microbiology, Biotechnology and other biological sciences. The contents of the book are designed to meet primarily the requirements of UG and PG students of Indian Universities. Attempt has been made to cover the entire syllabus and offer a text that is concise, simple and easy to understand.

The book contains chapters covering specific taxonomy and classification of viruses of vertebrates, invertebrates, plants, bacteria, algae and fungi. Bacterial virology covers important phages like lambda, M13, T-series of bacteriophages, their properties, and replication strategies. Plant virology comprises extensive taxonomical classification and properties of plant viruses, detailed description of significant viruses like tobacco mosaic virus, cauliflower mosaic virus, cotton leaf virus, rice stripe virus, potato virus-X, virus transmission, plant diseases, and their control measures. Clinical virology is updated with latest information for viruses of current importance and dealt in with detail; finally an extensive coverage of antiviral drugs and viral vaccines is imparted in this textbook.

I am sure that this book will be of immense use for undergraduate and postgraduate students in Microbiology, Biotechnology and Microbiology.

He has also tried to give up-to-date information on Virology incorporating new approaches and ideas emerging and expanding rapidly. I hope this book will receive attention and appreciation from students, teachers and healthcare professionals. I congratulate the author, Dr. Saravanan, for his commendable efforts and hope that students will use the book and benefit from it.

Prof. Dr. S. Ananthan Ph.D. (FoMedicine)
M.D.(Alt.Med.), F.I.M.S.A., F.A.B.M.S
Formerly Professor of Microbiology
University of Madras

PREFACE

The concept of viruses as a natural phenomenon separate from other infectious agents is more than 100 years old. Recognition of the fact that many diverse diseases of plants and animals can be attributed to this newly recognized type of agent came more lately. So did the recognition of phages that kill bacteria. In more recent years, the appreciation that practically all living species may have viruses associated with them distinguish the widespread prevalence of these agents and their potential importance in nature. Although not all viruses are pathogenic in their host, several have raised great concern in the present century because of the serious epidemic threats in humans like influenza, poliomyelitis, herpes, AIDS and childhood diarrhoea. Most texts do not attempt to cover the entire field of virology because of the breadth of this subject. The present attempt was made to fill this gap as a prerequisite to the use of this book.

The chapters in this book describe the properties of the major virus families of current importance starting with bacterial virology, then to plants, humans and finally attempts to provide latest information on antiviral drugs and viral vaccines. The virus families of plants and humans are discussed in a systematic manner. Properties of viruses, major viral diseases, mechanism of transmission, prevention and control measures are dealt in detail in the text. As a means of fine-tuning the reader's thought, a set of short questions are given at the end of each chapter for needful analysis of the subject. The purpose of this book is to introduce the reader to a wide informative data on viruses infecting bacteria, plants and animals, and to meet the requirements of a complete material on Virology.

In a field where knowledge is still rapidly expanding and interpretations are often changing, references to information in current scientific publications, which are not yet necessarily substantiated, should be more useful to the curious and interested student than a detailed description of possibly obsolete experimental data and conclusions. I trust that this textbook of Virology will be of value to readers particularly students of Microbiology and allied sciences following this constantly evolving and challenging field of science. I would appreciate and welcome suggestions from the readers for improvement of the text content.

I specially thank for the encouragement and advice on this book given by Professor Dr. S. Ananthan, formerly Professor of Microbiology, University of Madras, Chennai, and recipient of the prestigious Dr. B. C. Roy award for his outstanding contributions in the field of diarrhoeal diseases in children and AIDS patients. My thanks are also due to the publishers, who allowed my impediments to cope up with their time during final stages of production of the book.

Dr. P. Saravanan

CONTENTS

1 INTRODUCTION

Virus (derived from the Latin word *virus*, meaning "poison"), straddles the definition of life. They are intermediates to supramolecular complexes and very simple biological entities. Viruses do exhibit some of the activities that are common to organic life, despite having many dissimilarities.

Viruses exist in two distinct states. When not in contact with a suitable host cell, they remain entirely dormant. During this time there are no internal biological activities occurring within the virus, and in essence it is no more than a static organic particle. During this simple, clearly non-living state, viruses are referred to as virions. Virions can remain in this dormant state for extended periods of time, waiting in patience for a suitable host. When the virion comes into contact with an appropriate host, it becomes active and is then referred to as a virus. It now displays the properties characteristic of living organisms, such as reacting to its environment and directing its efforts toward self-replication. Viruses are composed of only one type of nucleic acid (either DNA or RNA, but never both) surrounded by a protective protein coat. Further they are subcategorized into dsDNA, ssDNA, dsRNA and ssRNA genome containing viruses. Some animal viruses have a phospholipid envelope surrounding the virus.

The discovery of viruses was made initially possible by the revolution in microbiology led by Louis Pasteur and his concept was refined by Robert Koch and Joseph Lister at the latter half of the 19th century. These scientists along with others discovered that microscopic organisms produced disease and the presence of the microbe could be defined by several criteria such as their retention on a Pasteur–Chamberlain filter, visualization by light microscopy and growth on agars and in broths. However, a subset of diseases failed these experimental tests and new classes of organisms were uncovered. Specifically, independent observations by Ivanovsky (1892, 1899) and Beijerinck (1898) of infected tobacco leaves, and by Loeffler and Frosch (1898) of the foot–and–mouth disease of cattle, demonstrated the presence of an infectious agent that passed through the porcelain filter to the bottom of the flask. These agents were not detected by light microscopy, and failed to grow in agar/broth cultures (Table 1.1). Indeed it was not until 40 years later, in the 1930s, that tobacco mosaic virus was observed by electron microscopy and nearly 50 years before that poliovirus could be reproducibly grown in tissue culture cells. The ability to grow viruses in cultured cells was the technological breakthrough that freed virologists from the need to pass viruses in animals and allowed sufficient recovery of viruses to allow the birth and development of biochemical virology in the 1960s, and its expansion to the molecular biology of viruses in the 1970s and 1980s.

With sufficient amount of viruses on hand, the study on viral structure, function and genomic organization followed, leading to insights into replication, transcription, translation, latency and integration and the beginning of our understanding of how viruses transform cells to a cancerous state or persisted among animals. In view of their lesser structural complexity, viruses became useful instruments for the study of genetics, cell biology, structural biology and biochemistry, allowing the study of many basic parameters of life and development. In addition, cell culture permitted the development of large-scale production of vaccines to prevent viral infections for instance, paralytic poliomyelitis, measles, rubella and hepatitis B virus (HBV) infections. These achievements led to the eradication of viral diseases that had killed over 400 million people in the 20th century alone, over a 4-fold greater than those killed by all the wars of this century. An estimated 12%–20% of human cancers are caused by viruses, and so the strategy utilized to prevent liver cancer caused by HBV will surely be applied to other cancers in the 21st century.

Table 1.1   Selected milestones in virology

Date(s)	Virologist(s)	Discovery
1798	E. Jenner	Small pox vaccine developed
1885	L. Pasteur	Rabies vaccine developed
1892, 1898	D. I. Ivanovski and M. Beijerinck	First demonstrations of a filterable plant virus: tobacco mosaic virus
1898	F. J. Loeffler and P. Frosch	First demonstration of a filterable animal virus: foot-and-mouth disease virus
1900	W. Reed and co-workers	Discovery of the cause of Yellow fever
1901	W. Reed and co-workers	First use of consent form for human clinical investigation (Yellow Fever Virus Commission)
1903	P. Remlinger and Riffat–Bey	Rabies virus
1904–1908	V. Ellermann and O. Bang, H. Vallee and H. Carre	First demonstration of a leukemia-causing virus, retrovirus
1907	P. M. Asburn and C. F. Craig	Dengue fever virus
1908–1911	C. Von Pirquet, J. Goldberger and J. F. Anderson	First report of virus causing immunosuppression: measles virus
1909	K. Landsteiner and E. Popper, S. Flexner and P. A. Lewis	Isolation of poliomyelitis virus
1911	P. Rous	First demonstration of a solid tumour virus: Rous sarcoma virus
1911	J. Goldberger and J. F. Anderson	Measles virus
1915	F. W. Twort	Discovery of bacteriophages

(Contd.)

Table 1.1   Contd.

Date(s)	Virologist(s)	Discovery
1917	F. d'Herelle	Bacteriophages, plaque assay
1923–1928	A. Carrel, H. Maitland and M. Maitland	Tissue culture of embryo explants and first tissue culture cultivation of virus: Rous virus, vaccinia virus
1931	J. Furth	Use of mice as hosts for viruses
1931	A. Woodruff and E. Goodpasteur	Use of embryonated hen's egg as a host for viruses
1933	W. Smith, C. H. Andrews and P. P. Laidlaw	Isolation of human influenza virus
1933	R. E. Shope	Rabbit papilloma virus: first DNA tumour virus
1934	C. D. Johnson and E. W. Goodpasture	Mumps virus
1935	Wendell M. Stanley	Crystallized TMV
1936	P. Rous, J. Beard	Rabbit papilloma virus that induces carcinoma in a different species
1938	Y.Hiroals and S. Tasaka	Rubella virus
1939	G. A. Kausche, E. Pfankuch, E. Ruska	First electron micrograph of a virus: tobacco mosaic virus
1946	W. Stanley, J. Summer, J. Northrop	Preparation of a viral protein in pure form: tobacco mosaic virus
1948–1955	J. F. Enders, P. C. Robbins, T. H. Weller	Routine use of tissue culture to grow and study viruses (polio virus)
	H. Eagle	Development of optimal media for growing cells
1937–1951	M. Theiler, H. Smith	Development of 17D strain for human yellow fever vaccine: made in animal and embryonic cell cultures

(Contd.)

Table 1.1 Contd.

Date(s)	Virologist(s)	Discovery
1951	Alfred D. Hershey and Martha Chase	Phage DNA enters cell
1953	W. P. Rowe	Adenoviruses discovered
1955	F. L. Schaffer and C. E. Schwerdt	Poliovirus crystallized
1954–61	J. Salk, A. Sabin, H. Koprowski, J. Enders, S. Katz, S. Krugman	Development of poliomyelitis virus vaccine and measles virus vaccine: made in tissue culture
1957	A. Isaacs, J. Lindenmann	Discovery of interferon
1950–70	S. Luria, M. Delbriick, J.Monod, E. Wollman, A. Hershey *et al.*	Quantitative plaque assay, origins of molecular biology and molecular virology
1950,	B. Sigurdsson, B. Blumberg,	Persistent, latent and slow virus infections
1960	C. Gajdusek	Prions
1959	L. Kilham and L. J. Olivier	Parvoviruses discovered
1967	T. O. Diener and W. B. Raymer	Nature of viroids elucidated
1970–80	R. Zinkernagel and P. Doherty, M. Oldstone B. Fields, B. Moss *et al.*	Major histocompatibility restriction and cytotoxic T-lymphocytes, immune mediated viral diseases, molecular pathogenesis
1969–76	R. Heubner, P. Vogt, M. Bishop, H. Varmus *et al.*	Oncogenes

(Contd.)

Table 1.1   Contd.

Date(s)	Virologist(s)	Discovery
1977	World Health Organization—Many health workers and virologists	Eradication of smallpox as a disease that killed over 300 million people in the twentieth century.
1977	F. Sanger and coworkers	DNA sequence of $\phi \times 174$ Bacteriophage determined
1978–85	S. Harrison, A. Olson, J. Hogle, M. Rossman, R. Rueckert	First atomic structure of a plant (tomato bush stunt) and animal (poliomyelitis, rhinovirus) virus
1979	D. Lane, L. Crawford, D. Linzer, A. Levine	SV40T—antigen—p53, virus host cell interaction—tumour suppressors
1981	J. Skehel, D. Wiley, I. Wilson *et al.*	Structure/function of influenza virus haemagglutinin. First atomic structure of a glycoprotein
1981–84	R. Gallo, F. Barre-Sinoussi, L. Montagnier, Y. Hinuma	First human retrovirus: Human Immunodeficiency Virus (HIV), cause of AIDS
1984–2000	M. Hilleman *et al.*	First molecular recombinant virus vaccine: hepatitis B virus. First vaccine to successfully treat cancer: hepatitis B virus-induced liver cancer
2000–2015	World Health Organization—Many health workers and virologists	Planned radication / elimination of poliomyelitis virus (by 2002–2005) and measles virus (by 2015)

Perhaps the first written record of a virus infection consists of a heiroglyph from Memphis, the capital of ancient Egypt, drawn in approximately 3700 BC, which depicts a temple priest called Ruma showing typical clinical signs of paralytic poliomyelitis (figure 1.1).

Figure 1.1   A heiroglyph from Egypt (3700 BC) depicting a temple priest showing typical clinical signs of paralytic poliomyelitis.

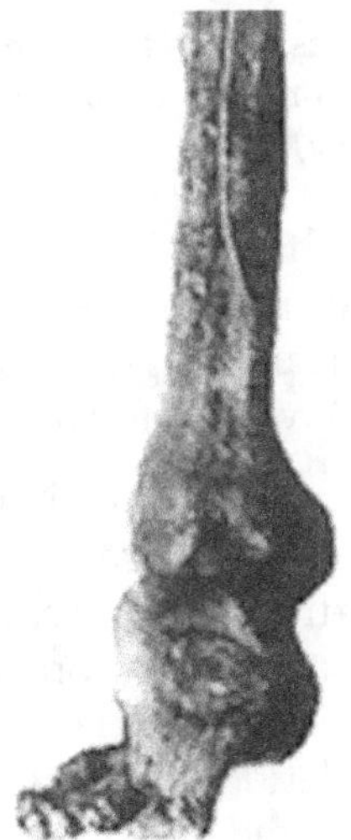

Figure 1.2   Skeleton of foot of excavated mummy showing foot rigidly extended by a horse's hoof.

The Pharaoh Siptah ruled Egypt from 1200–1193 BC where suddenly he died at an age of 20. His mummified body, laid undisturbed in his tomb in the Valley of the Kings until 1905. But later when the tomb was excavated, the mummy revealed that his left leg was withered and his foot was rigidly extended like a horse's hoof—classic paralytic poliomyelitis (Figure 1.2).

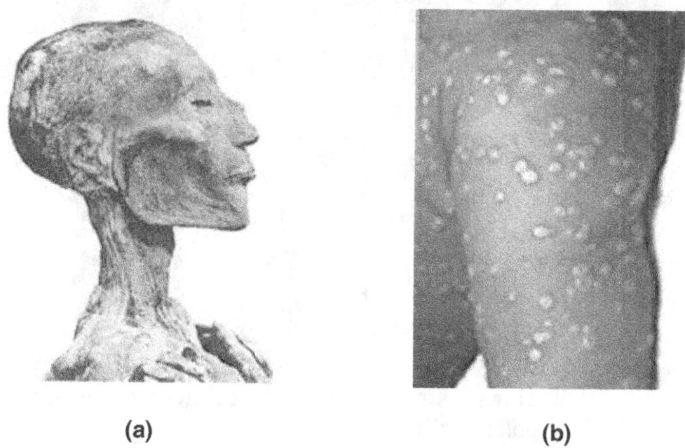

(a)   (b)

Figure 1.3   The pox lesions on the face of an excavated mummy (a), in comparison to the pustular lesions on the thigh of a patient (b).

In addition, the Pharaoh Ramses V, who died in 1196 BC, is believed to have succumbed to smallpox. The pustular lesions on the face of the mummy (figure 1.3a) and those of more recent patients (figure 1.3b) were found to be the same when compared.

Smallpox was endemic in China by 1000 BC. In response, the practice of variolation was developed. Recognizing that survivors of smallpox outbreaks were protected from subsequent infection, variolation involved inhalation of the dried crusts from smallpox lesions like snuff, or in later modifications, inoculation of the pus from a lesion into a scratch on the forearm of a child.

On 14 May, 1796, Edward Jenner used cowpox-infected material obtained from the hand of Sarah Nemes, a milkmaid from his home village of Berkeley in Gloucestershire to successfully vaccinate 8-year-old James Phipps. On 1st July, 1796, Jenner challenged the boy by deliberately inoculating him with material from a real case of smallpox! He did not become infected.

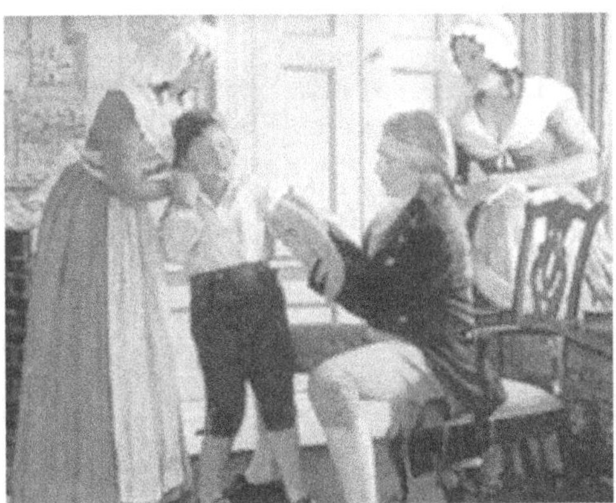

Figure 1.4 Inoculation of the pus from smallpox lesions into the scratch on the forearm of a child.

However, it was not until when Robert Koch and Louis Pasteur jointly proposed the "germ theory" of disease in the 1880s that the significance of these organisms became apparent.

ROBERT KOCH (1843–1910)

LOUIS PASTEUR (1822–1895)

Koch defined the four famous criteria now known as *Koch's postulates* which are still generally regarded as the proof that an infectious agent is responsible for a specific disease:

1. The agent must be present in every case of the disease.
2. The agent must be isolated from the host and grown in vitro.
3. The disease must be reproduced when a pure culture of the agent is inoculated into a healthy susceptible host.
4. The same agent must be recovered once again from the experimentally infected host.

Subsequently, Pasteur worked extensively on rabies, which he identified as being caused by a 'virus' (from the Latin for 'poison') but in spite of this, he did not discriminate between bacterial and other agents of disease.

On 12 February, 1892, **Dmitri Ivanovski**, a Russian botanist, presented a paper to the St. Petersburg Academy of Science which showed that extracts from diseased tobacco plants could transmit disease to other plants after passing them through ceramic filters fine enough to retain the smallest known bacteria. This is generally recognized as the beginning of **Virology**. Unfortunately, he did not fully realize the significance of these results.

DMITRI IVANOVSKI (1864–1920)   MARTINUS BEIJERINCK (1851–1931)

A few years later, in 1898, **Martinus Beijerinck** confirmed and extended Ivanovski's results on tobacco mosaic virus and was the first to develop the modern idea of the virus, which he referred to as contagium vivum fluidum ("soluble living germ").

Also in 1898, **Freidrich Loeffler and Paul Frosch** showed that a similar agent was responsible for foot-and-mouth disease in cattle. Thus, these new agents caused disease in animals as well as plants. In spite of these findings, there was resistance to the idea that these mysterious agents might have anything to do with human diseases. This view was finally dispelled by Landsteiner and Popper (1909), who showed that poliomyelitis was caused by a "filterable agent"— the first human disease to be recognized as having a viral cause.

FREIDRICH LOEFFER (1852–1915)

Frederick Twort (in 1915) and **Felix d'Herelle** (in 1917) were the first to recognize viruses which infect bacteria, which d'Herelle called bacteriophages (eaters of bacteria). In the 1930s and subsequent decades, pioneering virologists such as Luria, Delbruck and many others utilized these viruses as model systems to investigate many aspects of virology, including virus structure, genetics, replication, etc.

In 1881, Louis Pasteur began studies of rabies in animals. Over a number of years, he developed methods of producing attenuated virus preparations by progressively drying the spinal cords of rabbits experimentally infected with the agent, which when inoculated into animals, would protect from challenge with virulent virus. This was the first artificially produced virus vaccine.

FREDRICK TWORT (1877–1950)

FELIX D'HERELLE (1873–1949)

During the Spanish–American War of the late 19th century American deaths due to yellow fever were colossal. The disease also appeared to be spreading slowly northwards into the continental United States. Through experimental transmission to mice, in 1900 **Walter Reed** demonstrated that yellow fever was caused by a virus, spread by mosquitoes.

WALTER REED (1851–1902)

Reed's discovery eventually enabled **Max Theiler** (1937) to propagate the virus in chick embryos and successfully produced an attenuated vaccine—the 17D strain—which is still in use today. The success of this approach led many other investigators during the 1930s–1950s to develop animal systems to identify and propagate many other pathogenic viruses. Eukaryotic cells can be grown in vitro (tissue culture) and viruses can be propagated in these cultures, but these techniques are expensive and technically demanding.

MAX THEILER (1899–1972)

Some viruses will replicate in the living tissues of developing *embryonated hens eggs,* such as influenza virus. Egg-adapted strains of influenza virus replicate well in eggs and very high virus titers can be obtained. Embryonated eggs were first used to propagate viruses in the early decades of this century. This method has proved to be highly effective for the isolation and culture of many (but not all) viruses.

In recent years, an entirely new technology has been employed to study the effects of viruses on host organisms: the creation of transgenic animals and plants by means of the insertion of all or part of the virus genome into the DNA of the experimental organism, resulting in expression in the somatic cells (and sometimes in the cells of the germ line) of virus mRNA and proteins.

Cell culture began early this century with *whole organ cultures,* then progressed to methods involving individual cells, either *primary cell cultures* (somatic cells from an experimental animal or taken from a human patient which can be maintained for a short period in culture) or *immortalized cell lines*, which when given in appropriate conditions, continue to grow in culture indefinitely. In 1949, **Enders** and his colleagues were able to propagate poliovirus in primary human cell cultures. Renato Dulbecco in 1952 was the first to accurately quantify animal viruses using a *plaque assay*—dilutions of the virus are used to infect a cultured cell monolayer, which is then covered with soft-agar to restrict diffusion of the virus, resulting in localized cell killing and the appearance of plaques after the monolayer is stained. Counting the number of plaques directly determines the number of infectious virus particles applied to the plate.

Serological/Immunological Methods   In 1941, **Hirst** observed haemagglutination of red blood cells by influenza virus. This proved to be an important tool not only in the study of influenza, but also with several other groups of viruses, e.g. rubella virus. In the 1960s and subsequent years, many improved detection methods for viruses were developed. For example:

- Complement fixation test
- Radioimmunoassays
- Immunofluorescence (direct detection of virus antigens in infected cells or tissue)
- Enzyme Linked Immunosorbent Assays
- Radioimmune precipitation
- Western blot assays

These techniques are sensitive, quick and quantitative.

In 1975, Kohler and Milstein isolated the first *monoclonal antibodies* from clones of cells selected in vitro to produce an antibody of a single specificity directed against a particular antigenic target. This enabled virologists to look not only at the whole virus, but at specific regions—epitopes—of individual virus antigens. In recent years, this ability has greatly increased our understanding of the function of individual virus proteins. Monoclonal antibodies are also finding increasingly widespread application in other types of serological assays, e.g. ELISA, to increase their reproducibility, sensitivity and specificity.

ULTRASTRUCTURAL STUDIES

These can be considered under three areas:

1. Physical methods
2. Chemical methods
3. Electron microscopy

Physical measurements of virus particles began with the earliest determinations of their size by *filtration* through colloidal membranes with various pore sizes in the 1930s. Experiments of this sort led to the first (rather inaccurate) estimates of the size of virus particles. The accuracy of these was improved by studies of the *sedimentation properties* of viruses in ultracentrifuges in the 1960s. The physical properties of viruses can also be determined by *spectroscopy*, using both ultraviolet light to examine the nucleic acid content of the particle and visible light to determine its light scattering properties.

Electrophoresis of intact virus particles has yielded some limited information, but electrophoretic analysis of individual virion proteins by gel electrophoresis and also particularly of nucleic acid genomes, has been far more valuable in the diagnosis of viruses containing segmented genome such as rotavirus and influenza virus.

However, by far the most important method for the elucidation of virus structures has been the study of *X-ray diffraction* by crystalline forms of purified virus (crystallography). This technique permits determination of the structure of virions at an atomic level. The complete structures of many viruses, representative of many of the major groups, have now been determined at a resolution of a few angstroms.

This has advanced our understanding of the functions of the viral particle considerably.

However, many viruses are resistant to this type of investigation:

1. The virus must first be purified without loss of structural integrity. Therefore, adequate quantities of the virus and a subsequent purification method must be available.
2. The purified virus must be able to form crystals big enough to cause significant diffraction of the radiation source.

For some viruses, this is relatively straightforward and crystals big enough to see with the naked eye and diffract strongly are easily formed. This is particularly true for a **number of plant viruses, such as** *tobacco mosaic virus* (TMV) (which was first crystallized by Stanley in 1935) and *turnip yellow mosaic virus* (TYMV), the structures of which were among the first to be determined during the 1950s. It is significant that these two viruses represent the two fundamental types of virus particle; helical in the case of TMV and isometric for TYMV.

Nuclear magnetic resonance (NMR) is increasingly being used to determine the atomic structure of all kinds of molecules, including proteins and nucleic acids. The limitation of this method is that only relatively small molecules can be analysed. At present, the upper size limit for this technique restricts its use to molecules with a molecular weight of less than about 30–40,000—considerably less than even the smallest virus particles. Nevertheless, this method may well prove to be of great value in the future, certainly for examining isolated viral proteins, if not for intact virions.

Chemical investigation can be used to determine not only the overall composition of viruses and the nature of the nucleic acid which comprises the virus genome, but also the construction of the particle and the way in which individual components relate to each other in the capsid. The reagents used to denature virus capsids can indicate the basis of the stable interactions between its components:

- Proteins bound together by *electrostatic* interactions can be eluted by addition of ionic salts or alteration of pH.
- Proteins bound by non-ionic, *hydrophobic* interactions can be eluted by reagents such as urea.
- Proteins which interact with *lipid components* can be eluted by non-ionic detergents or organic solvents.

In addition to its fundamental structure, *progressive denaturation* can be used to observe alteration or loss of antigenic sites on the surface of particles and in this way, a picture of the physical state of the particle can be built up. Proteins exposed on the surface of viruses can be labelled with various compounds (e.g. iodination) to indicate those parts of the protein which are exposed and are protected inside the particle or by lipid membranes. *Cross-linking reagents* such as psoralens, or newer synthetic reagents with side arms of specific lengths are used to determine the spatial relationship of proteins and nucleic acids in intact viruses.

In 1931, **Ernst Ruska** constructed the first electron microscope. Electron microscopes overcome the fundamental limitation of light microscopes from the virologist's point of view, i.e., the inability to resolve individual virus particles due to physical constraints caused by the wavelength of visible light illumination. During subsequent years, techniques were developed which allowed the direct examination of viruses at magnifications of over 100,000X.

There are two fundamental types of electron microscopes, viz., the transmission *electron microscope (TEM)* and *scanning electron microscope (SEM)*.

Although beautiful images with three-dimensional appearance are produced by the SEM, for practical investigations of virus structure, the higher magnifications achievable with the TEM have proved to be of greater value. Two fundamental types of information can be obtained by electron microscopic images of viruses.

- The absolute number of virus particles present in any preparation (total count).
- The appearance/structure of the particles.

MOLECULAR BIOLOGY

Viral infection has often been used to probe the working of "normal" (i.e., uninfected) cells, e.g. macromolecular synthesis. This is true, for example, in the application of bacteriophages in bacterial genetics and also in many instances where eukaryotic viruses have revealed fundamental information about the cell biology and genomic organization of higher organisms. It is now possible to determine the nucleotide sequences of entire virus genomes, over a thousand of which have now been completed. This nucleic acid-centered technology also allowed advances in detection of viruses and viral infections involving nucleic acid hybridization techniques. A *hybridization probe*, labelled to facilitate detection, is allowed to react with a mixture of nucleic acids. The specific interaction of the probe sequence with complementary virus-encoded sequences, to which it binds by *hydrogen-bond formation* between the complementary base pairs, reveals the presence of the viral genetic material.

Because of the essentially digital nature of the nucleotide sequences, computers have been found to be the ideal means of storing and processing this information. Moreover, they are capable of predicting, with reasonable accuracy, the protein sequences (and detecting other functional regions) from nucleotide sequences. Vast international databases of nucleotide and protein sequence information have now been compiled, and these can be rapidly consulted with computers to compare newly determined sequences with those whose function may have been studied in great detail.

THE ORIGIN OF VIRUSES

There are (at least!) three theories which seek to explain the origin of viruses:

1. *Regressive evolution*   Viruses are degenerate life forms which have lost many functions unlike other organisms and have only retained the genetic information essential to their parasitic way of life.

2. *Cellular origin*   Viruses are subcellular, functional assemblies of macromolecules which have escaped their origins inside cells.

3. *Independent entities*   Viruses evolved on a parallel course to cellular organisms from the self-replicating molecules believed to have existed in the primitive prebiotic "RNA world".

VIRUS TAXONOMY AND CLASSIFICATION

The International Committee on Taxonomy of Viruses (ICTV)

In 1966 the International Committee on Nomenclature of Viruses (ICNV) was established at the International Congress of Microbiology in Moscow. In the later years by 1973, the ICNV became the International Committee on Taxonomy of Viruses (ICTV). Today, the ICTV operates under the auspices of the Virology Division of the International Union of Microbiological Societies. The ICTV has six subcommittees, 45 study groups, and over 400 participating virologists.

The ICTV database (ICTVdB)   The number of viruses occupying geographic and/or host niches as pathogens or silent passengers of humans, animals, plants, invertebrates, protozoa, fungi and bacteria are very large. Today, the ICTV recognizes more than 3,600 virus species. Speciality groups keep track of far more viruses, virus strains and subtypes, each having particular health or economic distinction and importance. It has been estimated that more than 30,000 viruses, virus strains and subtypes are being tracked in various speciality laboratories, reference centres and culture collections communicating with the WHO, FAO and other international agencies. Further, the development of the viral quasi species concept, with its prediction of rapid evolution of variants that may become fixed in nature as new species, portends future needs to track even more viral entities.

A major goal of the ICTV is to design, build and make available to all virologists, worldwide, a universal virus database, the ICTVdB. This database will encompass data that are now used in developing and managing the universal system for virus taxonomy. The ICTVdB will first describe viruses down to the species level, in keeping the level of responsibility of the ICTV, but it will then go further, interfacing with the databases of international speciality groups which are cataloguing data down to subspecies, strain, variant and isolate levels, that is, levels important in medicine, agriculture and other scholarly fields.

There are several virus databases in operation in the world which will be integrated into the ICTVdB, including:

1. The plant virus database operated at the Australian National University in Canberra, Australia (the VIDE Project; A. J. Gibbs, personal communication, 1994),

2. The veterinary virus database operated by the CSIRO/ Australian Animal Health Laboratory in Geelong, Australia (the VIREF Project; A. Della Porta, personal communication, 1994),

3. An arbovirus database operated for the American Committee on Arthropod-borne Viruses (ACAV) by the Centers for Disease Control in Fort Collins, Colorado, USA. Additionally, a human virus/human disease database, in the planning stage in the Division of Viral and Rickettsial Diseases, Centers for Disease Control in Atlanta, Georgia, USA, will be integrated into the ICTVdB. The most advanced of these databases is the

VIDE (Virus Identification Data Exchange) project on plant viruses. This project, centered in Canberra, Australia, interfaces with Horticulture Research International (Littlehampton, U.K.) and C.A.B. International (Wallingford, U.K.), and involves a worldwide network of more than 200 collaborating plant virologists. The VIDE database now contains 569 characters for more than 890 plant virus species in 55 genera. Using these databases as a foundation, the ICTV has laid out a plan to develop the ICTVdB over the next 10 years.

The Universal System of Virus Taxonomy

Today, there is a sense that a significant fraction of all existing viruses of humans, domestic animals and economically important plants have already been isolated and entered into the taxonomic system. This sense is based upon the lack of frequency in the recent year discoveries of viruses that do not fit into the present taxa. Of course, this sense does not extend to the viruses infecting the myriad of other species populating the Earth. This present sense of the diversity of the viruses, however imperfect, does point once again to the need for a universal, usable taxonomic system—a system to keep track of the large number of different viruses being isolated and studied throughout the world; a system to tie viral characteristics to virus names. The present universal system of virus taxonomy is useful and usable. It is set arbitrarily at hierarchical levels of order, family, subfamily, genus and species. Lower hierarchical levels, such as subspecies, strain, variant, etc. are established by international speciality groups and by culture collections.

Virus orders   Virus orders represent groupings of families of viruses that share common characteristics and are distinct from other orders and families. Virus orders are designated by names with the suffix -virales. To date, one order has been approved by the ICTV, the order Mononegavirales, comprising the families Paramyxoviridae, Rhabdoviridae and Filoviridae. It is ICTV's intention to move slowly in the approval of orders, limiting use to those instances where there is good evidence of phylogenetic relationship among the viruses of member families.

Virus families and subfamilies   Virus families represent groupings of genera of viruses that share common characteristics and are distinct from the member viruses of other families. Virus families are designated by names with the suffix -viridae. Despite concerns about the arbitrariness of early criteria for creating these taxa, most of the original families have stood the test of time and are still intact. This level in the taxonomic hierarchy now seems stable, and, indeed, is the benchmark of the entire universal system of taxonomy. Most of the families of viruses have distinct virion morphology, genome structure, and/or strategies of replication, indicating phylogenetic independence or great phylogenetic separation. At the same time, the virus family is being recognized as a taxon uniting viruses with a common, even if distant phylogeny. In four families, namely the families Poxviridae, Herpesviridae, Parvoviridae and Paramyxoviridae, subfamilies have been introduced to allow for a more complex hierarchy of taxa, in keeping with the apparent intrinsic complexity of the relationships among member viruses. Subfamilies are designated by terms with the suffix -virinae.

Virus genera   Virus genera represent groupings of species of viruses that share common characteristics and are distinct from the member viruses of other genera. Virus genera are designated by terms with the suffix -virus. This hierarchical level of taxa also seems stable and in many cases may be considered a benchmark for setting definitions of other taxa, especially species. The criteria used for creating genera differ from family to family. As more viruses are discovered and studied, there is pressure in many families to use smaller and smaller genetic, structural or other differences to create new genera. Since evidence of common phylogeny has entered the definition of many families, it is logical that even more such evidence will become the basis for defining genera. In fact, it might be said that in the future, genera will not stand where evidence is obtained of distinct phylogenies among member species.

Virus species   The species taxon has always been regarded as the most important hierarchical level in classification, but has been proved to be the most difficult to deal with viruses. After years of controversy, in 1991, the ICTV accepted the definition of a virus species proposed by van Regenmortel (1990), as follows: "A virus species is defined as a polythetic class of viruses that constitutes a replicating lineage and occupies a particular ecological niche." Members of a polythetic class are defined by more than one property and no single property is essential or necessary. One major advantage in this definition is that it can accommodate the inherent variability of viruses and it does not depend on the existence of a single unique characteristic. Similarly, it can accommodate the different traditions of virologists working in different areas of virology.

The ICTV study groups are now determining the specific properties to be used to define species in the taxon for which they are responsible. It seems clear that the species term will eventually be defined somewhat similarly to the term *virus*— although in some cases the term virus matches best with subspecies, strain or even variant. Just as the term virus is defined differently in different virus families, so is the species defined differently, in some cases with emphasis on genome properties, and in others on structural, physico chemical or serological properties. Some viruses have already been designated as species, for example: Sindbis virus, Newcastle disease virus, poliovirus 1, vaccinia virus, Fiji disease virus, tomato spotted wilt virus. These examples, however, do not reflect the difficulty that is being encountered in deciding whether a particular virus should be designated as a species, subspecies, strain or variant.

VIRUS NOMENCLATURE

Usage of Formal Taxonomic Nomenclature

In formal taxonomic usage, the first letters of virus family, subfamily and genus names are capitalized and the terms are printed in italics (underlined when typewritten). Species designations are not capitalized (unless they are derived from a place name or a host family or genus name), nor are they italicized. In formal usage, the name of the taxon should precede the term for the taxonomic unit; for example: "the family *Paramyxoviridae*", "the genus *Morbillivirus*." Furthermore, it was decided years ago that virus nomenclature would not involve the use of latinized binomial terms. For example, terms such as *Flavivirus fabricis*, *Orthopoxvirus variolae* and *Herpesvirus varicellae*, which were used at one time, have been abandoned. The following represent examples of full formal taxonomic terminology.

1. Family *Poxviridae*, subfamily *Chordopoxvirinae*, genus *Orthopoxvirus*, vaccinia virus.

2. Family *Herpesviridae*, subfamily *Alphaherpesvirinae*, genus *Simplexvirus*, human herpes virus 2 (herpes simplex virus 2).

3. Family *Picornaviridae*, genus *Enterovirus*, poliovirus 1.

4. Order *Mononegavirales*, family *Rhabdoviridae*, genus *Lyssavirus*, rabies virus.

5. Family *Bunyaviridae*, genus *Tospovirus*, tomato spotted wilt virus.

6. Family *Bromoviridae*, genus *Bromovirus*, brome mosaic virus.

7. Genus *Sobemovirus*, Southern bean mosaic virus.

8. Family *Totiviridae*, genus *Totivirus*, *Saccharomyces cerevisiae* virus L-A.

9. Family *Tectiviridae*, genus *Tectivirus*, enterobacteria phage PRD1.

10. Family *Plasmaviridae*, genus *Plasmavirus*, Acholeplasma phage L2.

Vernacular Usage of Virus Nomenclature

In informal vernacular usage, virus family, subfamily, genus and species names are written in lower case Roman script; they are not capitalized, nor are they printed in italics or underlined. In informal usage, the name of the taxon should not include the formal suffix, and the name of the taxon should follow the term for the taxonomic unit; for example, "the picornavirus family," "the enterovirus genus."

The use of vernacular terms for virus taxonomic units and virus names should not lead to unnecessary ambiguity or loss of precision in virus identification. The formal family, subfamily and genus terms and standard ICTV vernacular species terms, rather than any synonyms or transliterations, should be used as the basis for choosing vernacular terms.

One particular source of ambiguity in vernacular nomenclature lies in the common use of the same root terms in formal family and genus names. Imprecision stems from not being able to easily identify in vernacular usage which hierarchical level is being cited. For example, the vernacular name "paramyxovirus" might refer to the family *Paramyxoviridae*, the genus *Paramyxovirus*, or one of the species in the genus *Paramyxovirus*, such as one of the human parainfluenza viruses. Some virologists have suggested that this problem be solved by renaming taxa so that the same root term is never used at multiple hierarchical levels; however, there is no consensus for this, in fact, as plant virus taxonomy switches away from groups and towards families and genera, this problem will be exacerbated. The solution in vernacular usage is to avoid "jumping" hierarchical levels and to add taxon identification wherever needed.

For example, when citing the taxonomic placement of human parainfluenza virus 1, the term "paramyxovirus" should refer firstly to the genus, not the subfamily or family, and taxon identification should always be added: "human parainfluenza virus 1 is a member of the paramyxovirus genus," rather than "human parainfluenza virus 1 is a paramyxovirus." Most examples like this exemplify the advantage of switching, where necessary, into formal nomenclature usage: "human parainfluenza virus 1 is a species in the genus *Paramyxovirus*, family *Paramyxoviridae*." In this example, as is usually the case, adding the information that this virus is also a member of the subfamily *Paramyxovirinae* and the order *Mononegavirales* is unnecessary.

List of Criteria Demarcating Different Virus Taxa

I. Order

Common properties between several families including;

- Biochemical composition
- Virus replication strategy
- Particle structure (to some extent)
- General genome organization

II. Family

Common properties between several genera including;

- Biochemical composition
- Virus replication strategy
- Nature of the particle structure
- Genome organization

III. Genus

Common properties within a genus including;

- Virus replication strategy
- Genome size, organization and/or number of segments
- Sequence homologies (hybridization properties)
- Vector transmission

IV. Species

Common properties within a species including;

- Genome rearrangement
- Sequence homologies (hybridization properties)
- Serological relationships
- Vector transmission

- Host range
- Pathogenicity
- Tissue tropism
- Geographical distribution

Virus Family Descriptors Used in Virus Taxonomy

I. Virion properties

A. *Morphological properties of virions*
1. Virion size
2. Virion shape
3. Presence or absence of an envelope and peplomers
4. Capsomeric symmetry and structure

B. *Physical properties of virions*
1. Molecular mass of virions
2. Buoyant density of virions
3. Sedimentation coefficient
4. pH stability
5. Thermal stability
6. Cation (Mg^{++}, Mn^{++}) stability
7. Solvent stability
8. Detergent stability
9. Radiation stability

C. *Properties of genome*
1. Type of nucleic acid—DNA or RNA
2. Strandedness—single stranded or double stranded
3. Linear or circular
4. Sense—positive, negative or ambisense
5. Number of segments

6. Size of genome or genome segments
7. Presence or absence and type of 5´-terminal cap
8. Presence or absence of 5´-terminal covalently linked polypeptide
9. Presence or absence of 3´-terminal poly(A) tract (or other specific tract)
10. Nucleotide sequence comparisons

D. *Properties of proteins*

1. Number of proteins
2. Size of proteins
3. Functional activities of proteins (especially virion transcriptase, virion reverse transcriptase, virion hemagglutinin, virion neuraminidase and virion fusion protein)
4. Amino acid sequence comparisons

E. *Lipids*

1. Presence or absence of lipids
2. Nature of lipids

F. *Carbohydrates*

1. Presence or absence of carbohydrates
2. Nature of carbohydrates

II. Genome organization and replication

1. Genome organization
2. Strategy of replication of nucleic acid
3. Characteristics of transcription
4. Characteristics of translation and post-translational processing
5. Site of accumulation of virion proteins, site of assembly, site of maturation and release
6. Cytopathology, inclusion body formation

III. *Antigenic properties*

1. Serological relationships
2. Mapping epitopes

IV. *Biological properties*

1. Host range, natural and experimental
2. Pathogenicity, association with disease
3. Tissue tropisms, pathology, histopathology
4. Mode of transmission in nature
5. Vector relationships
6. Geographic distribution

I Families and Genera of Viruses that Infect Vertebrates

Group I: dsDNA Viruses

Order *Caudovirales* —Tailed Bacteriophages

Family (Subfamily)	Genus	Type Species	Hosts
Adenoviridae	*Mastadenovirus*	Human adenovirus C	Vertebrates
	Aviadenovirus	Fowl adenovirus A	Vertebrates
Asfarviridae	*Asfivirus*	African swine fever virus	Vertebrates
Herpesviridae	*Simplexvirus*	Human herpesvirus 1&2	Vertebrates
Alphaherpesvirinae	*Varicellovirus*	Human herpesvirus 3	Vertebrates
Betaherpesvirinae	*Cytomegalovirus*	Human herpesvirus 5	Vertebrates

(Contd.)

Group I : dsDNA Viruses (Contd.)

Family (Subfamily)	Genus	Type Species	Hosts
	Muromegalovirus	Murine herpesvirus 1	Vertebrates
	Roseolovirus	Human herpesvirus 6	Vertebrates
Gammaherpesvirinae	*Lymphocryptovirus*	Human herpesvirus 4	Vertebrates
	Rhadinovirus	Simian herpesvirus 2	Vertebrates
Iridoviridae	*Ranavirus*	Frog virus 3	Vertebrates
	Lymphocystivirus	Lymphocystis disease virus 1	Vertebrates
Polyomaviridae	*Polyomavirus*	Simian virus 40	Vertebrates
Papillomaviridae	*Papillomavirus*	Cotton tail rabbit papilloma virus	Vertebrates
Poxviridae	*Orthopoxvirus*	Vaccinia virus	Vertebrates
Chordopoxvirinae	*Parapoxvirus*	Orf virus	Vertebrates
	Avipoxvirus	Fowlpox virus	Vertebrates
	Capripoxvirus	Sheeppox virus	Vertebrates
	Leporipoxvirus	Myxoma virus	Vertebrates
	Suipoxvirus	Swinepox virus	Vertebrates
	Molluscipoxvirus	*Molluscum contagiosum* virus	Vertebrates
	Yatapoxvirus	Yaba monkey tumour virus	Vertebrates

Group II: ssDNA Viruses

Family (Subfamily)	Genus	Type Species	Hosts
Circoviridae	*Circovirus*	Porcine circovirus	Vertebrates
	Gyrovirus	Chicken anaemia virus	Vertebrates
Parvoviridae	*Parvovirus*	Mice minute virus	Vertebrates
Parvovirinae	*Erythrovirus*	B19 virus	Vertebrates
	Dependovirus	Adeno-associated virus 2	Vertebrates

Group III: dsRNA Viruses

Family (Subfamily)	Genus	Type Species	Hosts
Birnaviridae	*Aquabirnavirus*	Infectious pancreatic necrosis virus	Vertebrates
	Avibirnavirus	Infectious bursal disease virus	Vertebrates
Reoviridae	*Orthoreovirus*	Mammalian orthoreovirus	Vertebrates
	Orbivirus	Bluetongue virus	Vertebrates
	Rotavirus	Rotavirus A	Vertebrates
	Coltivirus	Colorado tick fever virus	Vertebrates
	Aquareovirus	Golden shiner virus	Vertebrates

Group IV: (+)-sense RNA Viruses

Order *Nidovirales* — "Nested" Viruses

Family (Subfamily)	Genus	Type Species	Hosts
Arteriviridae	*Arterivirus*	Equine arteritis virus	Vertebrates
Coronaviridae	*Coronavirus*	Infectious bronchitis virus	Vertebrates
	Torovirus	Equine totovirus	Vertebrates
Caliciviridae	*Lagovirus*	Rabbit haemorrhagic disease virus	Vertebrates
	Norwalk-like viruses	Norwalk virus	Vertebrates
	Sapporo-like viruses	Sapporo virus	Vertebrates
	Vesivirus	Swine vesicular exanthema virus	Vertebrates
Flaviviridae	*Flavivirus*	Yellow fever virus	Vertebrates
	Pestivirus	Bovine diarrhoea virus 1	Vertebrates
	Hepacivirus	Hepatitis C virus	Vertebrates
	Hepatitis E-like viruses	Hepatitis E virus	Vertebrates
Nodaviridae	*Betanodovirus*	Striped jack nervous necrosis virus	Vertebrates

(Contd.)

Group IV : (+)-sense RNA Viruses (Contd.)

Family (Subfamily)	Genus	Type Species	Hosts
Picornaviridae	*Enterovirus*	Poliovirus	Vertebrates
	Rhinovirus	Human rhinovirus A	Vertebrates
	Hepatovirus	Hepatitis A virus	Vertebrates
	Cardiovirus	Encephalomyo-carditis virus	Vertebrates
	Aphthovirus	Foot-and-mouth disease virus O	Vertebrates
	Parechovirus	Human parechovirus	Vertebrates
	Erbovirus	Equine rhinitis B virus	Vertebrates
	Kobuvirus	Aichi virus	Vertebrates
	Teschovirus	Porcine teschovirus	Vertebrates
Togaviridae	*Alphavirus*	Sindbis virus	Vertebrates
	Rubivirus	Rubella virus	Vertebrates

Group V: (–)-sense RNA Viruses
Order Mononegavirales

Family (Subfamily)	Genus	Type Species	Hosts
Bornaviridae	*Bornavirus*	Borna disease virus	Vertebrates
Filoviridae	*Marburg-like viruses*	Marburg virus	Vertebrates

(Contd.)

Group V : (–)-sense RNA Viruses (Contd.)

Family (Subfamily)	Genus	Type Species	Hosts
	Ebola–like viruses	Ebola virus	Vertebrates
Bunyaviridae	*Bunyavirus*	Bunyamwera virus	Vertebrates
	Hantavirus	Hantaan virus	Vertebrates
	Nairovirus	Nairobi sheep disease virus	Vertebrates
	Phlebovirus	Sandfly fever Sicilian virus	Vertebrates
	Deltavirus	Hepatitis delta virus	Vertebrates
Orthomyxoviridae	*Influenza A virus*	Influenza A virus	Vertebrates
	Influenza B virus	Influenza B virus	Vertebrates
	Influenza C virus	Influenza C virus	Vertebrates
	Thogotovirus	Thogoto virus	Vertebrates
Paramyxoviridae: Paramyxovirinae	*Respirovirus*	Sendai virus	Vertebrates
	Morbillivirus	Measles virus	Vertebrates
	Rubulavirus	Mumps virus	Vertebrates
Pneumovirinae	*Pneumovirus*	Human respiratory syncytial virus	Vertebrates
	Metapneumo virus	Turkey rhinotracheitis virus	Vertebrates

(Contd.)

Group V : (–)-sense RNA Viruses (Contd.)

Family (Subfamily)	Genus	Type Species	Hosts
Rhabdoviridae	*Vesiculovirus*	Vesicular stomatitis Indiana virus	Vertebrates
	Lyssavirus	Rabies virus	Vertebrates
	Ephemerovirus	Bovine ephemeral fever virus	Vertebrates
	Novirhabdovirus	Infectious haematopoetic necrosis virus	Vertebrates
Arenaviridae	*Arenavirus*	Lymphocytic choriomeningitis virus	Vertebrates

Group VI: RNA Reverse Transcribing Viruses

Family (Subfamily)	Genus	Type Species	Hosts
Retroviridae	*Alpharetrovirus*	Avian leukosis virus	Vertebrates
	Betaretrovirus	Mouse mammary tumour virus	Vertebrates
	Gammaretrovirus	Murine leukemia virus	Vertebrates
	Deltaretrovirus	Bovine leukemia virus	Vertebrates
	Epsilonretrovirus	Walley dermal sarcoma virus	Vertebrates

(Contd.)

Group VI : RNA reverse transcribing Viruses (Contd.)

Family (Subfamily)	Genus	Type Species	Hosts
	Lentivirus	Human immunodeficiency virus 1	Vertebrates
	Spumavirus	Human spumavirus	Vertebrates

Group VII: DNA Reverse Transcribing Viruses

Family (Subfamily)	Genus	Type Species	Hosts
Hepadnaviridae	*Orthohepadnavirus*	Hepatitis B virus	Vertebrates
	Avihepadnavirus	Duck hepatitis B virus	Vertebrates

II   Families and Genera of Viruses that infect Invertebrates, Bacteria, Plants, Fungi and Algae

Group I: dsDNA Viruses

Order *Caudovirales* —Tailed Bacteriophages

Family (Subfamily)	Genus	Type Species	Hosts
Myoviridae	*T4–like phages*	Enterobacteria phage T4	Bacteria
	P1–like phages	Enterobacteria phage P1	Bacteria
	P2–like phages	Enterobacteria phage P2	Bacteria

(Contd.)

Group I: dsDNA Viruses (Contd.)

Family (Subfamily)	Genus	Type Species	Hosts
	Mu–like phages	Enterobacteria phage Mu	Bacteria
	SP01–like phages	Bacillus phage SP01	Bacteria
	φH–like phages	Halobacterium virus AUH	Bacteria
Podoviridae	*T7–like phages*	Enterobacteria phage T7	Bacteria
	P22–like phages	Enterobacteria phage P22	Bacteria
	φ29–like phages	Bacillus phage φ29	Bacteria
Siphoviridae	*λ like phages*	Enterobacteria phage λ	Bacteria
	T1–like phages	Enterobacteria phage T1	Bacteria
	T5–like phages	Enterobacteria phage T5	Bacteria
	L5–like phages	Mycobacterium phage L5	Bacteria
	c2–like phages	Lactococcus phage c2	Bacteria
	ψM1–like phages	Methanobacterium virus ψM1	Bacteria

(Contd.)

Group I: dsDNA Viruses (Contd.)

Family (Subfamily)	Genus	Type Species	Hosts
Ascoviridae	*Ascovirus*	*Spodoptera frugiperda* ascovirus	Invertebrates
Baculoviridae	*Nucleopoly-hedrovirus*	*Autographa californica* nucleopoly-hedro virus	Invertebrates
	Granulovirus	*Cydia pomonella* granulovirus	Invertebrates
Corticoviridae	*Corticovirus*	Alteromonas phage PM2	Bacteria
Fuselloviridae	*Fusellovirus*	Sulfolobus virus SSV1	Archaea
Iridoviridae	*Iridovirus*	Invertebrate iridescent virus 6	Invertebrates
	Chloriridovirus	Invertebrate iridescent virus 3	Invertebrates
Lipothrixviridae	*Lipothrixvirus*	Thermo proteus virus 1	Archaea
Phycodnaviridae	*Chlorovirus*	*Paramecium bursaria* Chlorella virus 1	Algae
	Prasinovirus	*Micromonas pusilla* virus SP1	Algae

(Contd.)

Group I: dsDNA Viruses (Contd.)

Family (Subfamily)	Genus	Type Species	Hosts
	Prymnesiovirus	*Chryosochromomulina brevifilium* virus PW1	Algae
	Phaeovirus	*Ectocarpus siliculosus* virus1	Algae
Plasmaviridae	*Plasmavirus*	Acholeplasma phage L2	Mycoplasma
Polydnaviridae	*Ichnovirus*	*Campoletis sonorensis* ichnovirus	Invertebrates
	Bracovirus	*Cotesia melanoscela* brachovirus	Invertebrates
Poxviridae Entomopoxvirinae	*Entomopox virus A*	*Melolontha melolontha* entomopoxvirus	Invertebrates
	Entomopox virus B	*Amsacta moorei* entomopoxvirus	Invertebrates
	Entomopox virus C	*Chironomus luridus* entomopoxvirus	Invertebrates
	Rhizidovirus	Rhizidomyces virus	Fungi
Rudiviridae	*Rudivirus*	Sulfolobus virus SIRV1	Archaea
Tectiviridae	*Tectivirus*	Enterobacteria phage PRD1	Bacteria

Group II: ssDNA Viruses

Family (Subfamily)	Genus	Type Species	Hosts
Geminiviridae	*Mastrevirus*	Maize streak virus	Plants
	Curtovirus	Beet curly top virus	Plants
	Topocuvirus	Tomato pseudo-curly top virus	Plants
	Begomovirus	Bean golden mosaic virus	Plants
Inoviridae	*Inovirus*	Enterobacteria phage M13	Bacteria
	Plectrovirus	Acholeplasma phage MV–L51	Bacteria
Microviridae	*Microvirus*	Enterobacteria ϕ X174	Bacteria
	Spiromicrovirus	Spiroplasma phage 4	Spiroplasma
	Bdellomicrovirus	Bdellovibrio phage MAC1	Bacteria
	Chlamydiamicrovirus	Chlamydia phage 1	Bacteria
	Nanovirus	Subterranean clover stunt virus	Plants
Densovirinae	*Densovirus*	*Junonia coenia* densovirus	Invertebrates
	Iteravirus	*Bombyx mori* densovirus	Invertebrates
	Brevidensovirus	*Aedes aegypti* densovirus	Invertebrates

Group III: dsRNA Viruses

Family (Subfamily)	Genus	Type Species	Hosts
Birnaviridae	*Entomobirna-virus*	*Drosophila* X virus	Invertebrates
Cystoviridae	*Cystovirus*	*Pseudomonas* phage ϕ 6	Bacteria
Hypoviridae	*Hypovirus*	Cryphonectria hypovirus 1–EP713	Fungi
Partitiviridae	*Partitivirus*	*Gaeumarmomyces graminis* virus 019/6–A	Fungi
	Chrysovirus	*Penicillium chrysogenum* virus	Fungi
	Alphacryptovirus	White clover cryptic virus 1	Plants
	Betacryptovirus	White clover cryptic virus 2	Plants
Reoviridae	*Cypovirus*	Cypovirus 1	Invertebrates
	Fijivirus	Fiji disease virus	Plants
	Phytoreovirus	Wound tumour virus	Plants
	Oryzavirus	Rice ragged stunt virus	Plants
Totiviridae	*Totivirus*	*Saccharomyces cerevisiae* virus L–A	Fungi
	Giardiavirus	*Giardia lamblia* virus	Protozoa
	Leishmaniavirus	*Leishmania* RNA virus 1–1	Protozoa

Group IV: (+)-sense RNA Viruses
Order *Nidovirales* — "Nested" Viruses

Family (Subfamily)	Genus	Type Species	Hosts
Allexivirus		Shallot virus X	Plants
Barnaviridae	*Barnavirus*	Mushroom bacilliform virus	Fungi
	Benyvirus	Beet necrotic yellow vein virus	Plants
Bromoviridae	*Alfamovirus*	Alfalfa mosaic virus	Plants
	Bromovirus	Brome mosaic virus	Plants
	Cucumovirus	Cucumber mosaic virus	Plants
	Ilarvirus	Tobacco streak virus	Plants
	Oleavirus	Olive latent virus 2	Plants
	Capillovirus	Apple stem grooving virus	Plants
	Carlavirus	Carnation latent virus	Plants
Closteroviridae	*Closterovirus*	Beet yellow virus	Plants
	Idaeovirus	Raspberry bushy dwarf virus	Plants
Comoviridae	*Comovirus*	Cowpea mosaic virus	Plants
	Fabavirus	Broad bean wilt virus 1	Plants
	Nepovirus	Tobacco ringspot virus	Plants

(Contd.)

Group IV: (+)-sense RNA Viruses (Contd.)

Family (Subfamily)	Genus	Type Species	Hosts
	Cricket paralysis-like viruses	Cricket paralysis viruses	Invertebrates
Flaviviridae	*Foveavirus*	Apple stem pitting virus	Plants
	Furovirus	Soil-borne wheat mosaic virus	Plants
	Hordeivirus	Barley stripe mosaic virus	Plants
Leviviridae	*Levivirus*	Enterobacteria phage MS2	Bacteria
	Allolevivirus	Enterobacteria phage $\phi\beta$	Bacteria
Luteoviridae	*Luteovirus*	Cereal yellow dwarf virusβPAV	Plants
	Polerovirus	Potato leafroll virus	Plants
	Enamovirus	Pea enation mosaic virus-1	Plants
	Machlomovirus	Maize chlorotic mottle virus	Plants
	Marafivirus	Maize rayado fino virus	Plants
Narnaviridae	*Narnavirus*	*Saccharomyces cerevisiae* narnavirus 20S	Fungi
	Mitovirus	*Cryphonectria parasitica* mitovirus 1-NB631	Fungi

(Contd.)

Group IV: (+)-sense RNA Viruses (Contd.)

Family (Subfamily)	Genus	Type Species	Hosts
Nodaviridae	*Alphanodoavirus*	Nodamura virus	Invertebrates
	Pecluvirus	Peanut clump virus	Plants
	Ourmiavirus	Ourmia melon virus	Plants
	Pomovirus	Potato mop-top virus	Plants
	Potexvirus	Potato virus X	Plants
Potyviridae	*Potyvirus*	Potato virus Y	Plants
	Ipovirus	Sweet potato mild mottle virus	Plants
	Macluravirus	Maclura mosaic virus	Plants
	Rymovirus	Ryegrass mosaic virus	Plants
	Tritimovirus	Wheat streak mosaic virus	Plants
	Bymovirus	Barley yellow mosaic virus	Plants
Sequiviridae	*Sequivirus*	Parsnip yellow fleck virus	Plants
	Waikavirus	Rice tungro spherical virus	Plants
	Sobemovirus	Southern bean mosaic virus	Plants
Tetraviridae	*Betatetravirus*	*Nudaurelia capensis* β virus	Invertebrates
	Omegatetravirus	*Nudaurelia capensis* ω virus	Invertebrates

(Contd.)

Group IV: (+)-sense RNA Viruses (Contd.)

Family (Subfamily)	Genus	Type Species	Hosts
	Tobamovirus	Tobacco mosaic virus	Plants
	Tobravirus	Tobacco rattle virus	Plants
Tombusviridae	*Tombusvirus*	Tomato bushy stunt virus	Plants
	Avenavirus	Oat chlorotic stunt virus	Plants
	Aureusvirus	Pothos latent virus	Plants
	Carmovirus	Carnation mottle virus	Plants
	Dainthovirus	Carnation ring spot virus	Plants
	Machlomovirus	Maize chlorotic mottle virus	Plants
	Necrovirus	Tobacco necrosis virus	Plants
	Panicovirus	Panicum mosaic virus	Plants
	Trichovirus	Apple chlorotic leaf spot virus	Plants
	Tymovirus	Turnip yellow mosaic virus	Plants
	Umbravirus	Carrot mottle virus	Plants
	Vitivirus	Grapevine virus A	Plants

Group V: (–)-sense RNA Viruses
Order Mononegavirales

Family (Subfamily)	Genus	Type Species	Hosts
Rhabdoviridae	*Cytorhabdovirus*	Lettuce necrotic yellow virus	Plants
	Nucleorhabdovirus	Potato yellow dwarf virus	Plants
Bunyaviridae	*Tospovirus*	Tomato spotted wilt virus	Plants
	Ophiovirus	*Citrus psorosis* virus	Plants
	Tenuivirus	Rice stripe virus	Plants

Group VI: RNA Reverse Transcribing Viruses

Family (Subfamily)	Genus	Type Species	Hosts
Metaviridae	*Metavirus*	*Saccharomyces cerevisiae* Ty3 virus	Fungi
	Errantivirus	*Drosophila melanogaster* gypsy virus	Invertebrates
Pseudoviridae	*Pseudovirus*	*Saccharomyces cerevisiae* Ty1 virus	Invertebrates
	Hemivirus	*Drosophila melanogaster* copia virus	Invertebrates

Group VII: DNA Reverse Transcribing Viruses

Family (Subfamily)	Genus	Type Species	Hosts
Caulimoviridae	*Caulimovirus*	Cauliflower mosaic virus	Plants
	Badnavirus	Commelina yellow mottle virus	Plants

Subviral Agents: Viroids

Family (Subfamily)	Genus	Type Species	Hosts
Pospiviroidae	*Pospiviroid*	Potato spindle tuber viroid	Plants
	Hostuviroid	Hop stunt viroid	Plants
	Cocadviroid	Coconut cadang-cadang viroid	Plants
	Apscaviroid	Apple scar skin viroid	Plants
	Coleviroid	*Coleus blumei* viroid 1	Plants
Avsunviroidae	*Avsunviroid*	Avocado sunblotch viroid	Plants
	Pelamonviroid	Peach latent mosaic viroid	Plants

Subviral Agents: Satellites and Prions

Family (Subfamily)	Genus	Type Species	Hosts
Satellites			Invertebrates Plants Fungi
Prions			Vertebrates Fungi

REVIEW QUESTIONS

1. Write short notes on:

    a. Viral capsid

    b. Prions

    c. Virion and viroid

    d. Variolation

    e. Koch's postulates

    f. ELISA

    g. ICTV

    h. Any four DNA and RNA viruses

    i. TMV

    j. Edward Jenner

2. Give an account of the classification of viruses.

3. Detail the virus family descriptors used in virus taxonomy.

2   BACTERIAL VIROLOGY

Bacteriophages are bacterial viruses. Based on their nucleic acid composition, they are further grouped as dsDNA, ssDNA, dsRNA and ssRNA phages (ds=double stranded, ss=single stranded). They can exist outside the bacterial cell.

Virus enzymes like lysozyme are made by the phages to get out of the host. These enzymes degrade peptidoglycan. Some animal viruses make neuraminidase which breaks down glycolipids to let the virus get out of the cell. Retroviruses are ssRNA viruses that infect host cells and make a complementary copy of their genome in dsDNA which is inserted in a chromosome. These viruses code for an RNA-dependent DNA polymerase.

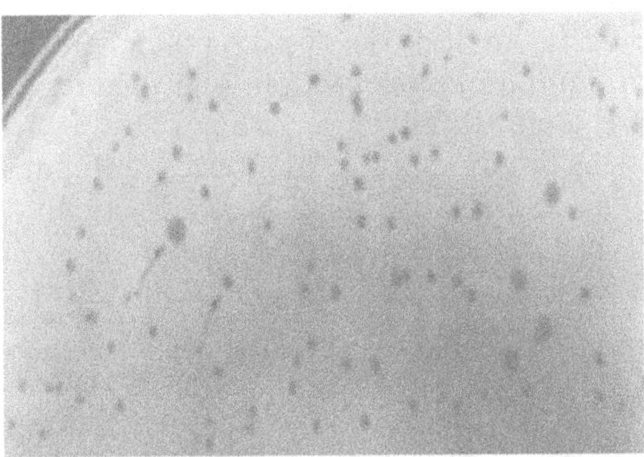

Figure 2.1   Plaque assay plate showing clearance of zones produced by bacteriophages on a lawn of bacteria.

The phages replicate on a lawn (thin layer) of bacteria on a petri plate. They lyse the bacteria around them (1–5 mm diameter) to form a *plaque* (figure 2.1). Phage numbers can be titrated by dilution plating in which serial 10-fold dilution of phages are made and plated on lawns of bacteria to measure their plaque numbers. Animal viruses do not form plaques very often. They have "cytopatheic effects" on cells, which can be visible. Animal viruses are usually enumerated with the electron microscope.

Temperate bacteriophage—lambda, the best known—can infect the cells with a lytic infection and make 100 copies. It can also recombine into the genome and be carried like any other gene by the bacterium. This is called *lysogeny*, the bacterium is said to be lysogenic when it carries a virus. The bacteriophage makes only the cI protein, a repressor that turns off all the bacteriophage genes. When the host DNA is damaged by UV light or by a chemical mutagen, the cI protein is destroyed by a protease and the bacteriophage goes through its life cycle.

Viroids are small, circular ssRNA—as few as 246 nucleotides—causing diseases in plants. There is neither a capsid nor any enzyme, but some secondary structure to the RNA is observed.

Prions are small pieces of proteins normally found in cells but have a unique conformation. They can make these normal proteins change conformation to form a prion. They are capable of causing bovine spongiform encephalopathy (mad cow disease,) kuru and Cruetzfeld–Jakob disease in humans.

TERMS USED IN DESCRIBING THE SIZE, SHAPE, AND TOPOGRAPHY OF VIRUSES AND THEIR COMPONENTS

Whenever we are dealing with aggregates of many molecules, usually of different nature, held together by secondary forces, such aggregates will be termed **particles** and described in **particle weights**. Thus, viruses have particle weights rather than molecular weights, usually a number $\times$ 106, since they are all in the millions (3 to 800 $\times$ 10^6). Their component nucleic acids, proteins, and occasionally lipids and others are molecules, since they consist of covalently linked atoms, and their molecular weights will be given. For example, TMV coat protein has a molecular weight of about 17,500, or a mass of 17,500 daltons (Da), but will be abbreviated as the 17.5 kDa protein ($k = 1000$).

All particle and molecular weight analyses of macromolecules are subject to considerable methodological error. Thus, to say that a virus coat is made up of two proteins of 42 kDa and 22 kDa means only that the method employed revealed the presence of only two proteins and that one is about twice as large as the other, the actual molecular weight of either probably being within the range of $\pm$ 20 percent of the given value. The only way to obtain the exact molecular weight of macromolecules is by complete structural analysis, amino acid sequencing in the case of proteins, and (deoxy) ribonucleotide sequencing in the case of nucleic acids. Another means of characterizing particles and large molecules is by their S value, derived from their sedimentation rate in the ultracentrifuge. These S values are reproducible under specific conditions, but they are not directly proportional to the mass of the particle or molecule. When discussing linear dimensions, we use in this text only nanometres (nm, 10^{-9} m, formerly called mm); we avoid the frequently used angstrom (Å), which equals 0.1 nm.

Two terms frequently used in virus studies are in vivo and in vitro. Because cell culture is generally done in glass dishes (*in vitro* in Latin), this term is often used for such studies. The term in vitro is also employed for biochemical studies of cellular components of disrupted cells. In vivo has, in general, been reserved for studies done in living hosts (e.g. animal or human subjects).

Most viruses are organized according to either of two structural principles, the **helical**, which gives them a rod like appearance, or the **isometric**, which gives them roughly spherical shapes. A group of terms has been proposed to identify the various structural components of viruses. The monomer of the protein that forms the viral coat is called the structural subunit. In isometric viruses a definite number of these subunits usually aggregate in a specific manner to form what is called a morphological unit, a body of a characteristic shape, generally discernible on electron micrographs; it is also called a capsomere.The orderly complex of many subunits or capsomeres with viral nucleic acid is the nucleocapsid, as contrasted with the empty capsid, the aggregated protein shell lacking nucleic acid. The complete virus particle, which in addition to the nucleocapsid may contain additional structural proteins and/or envelopes, is called the virion. The spikes protruding from the envelope may also be called peplomers. Most of these diverse structural features have been observed, at times in a very similar manner, in plant, animal, and bacterial viruses.

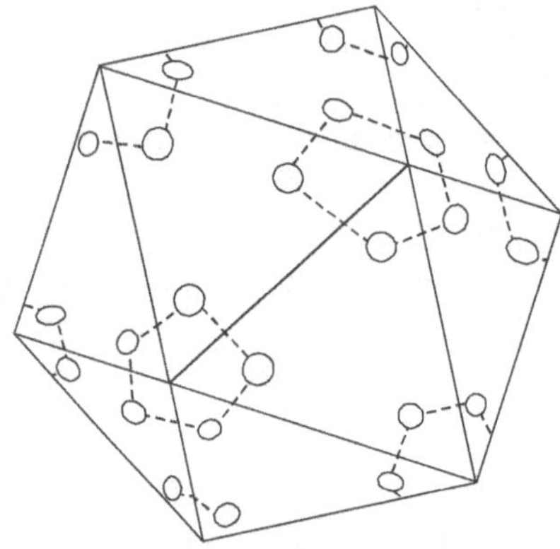

Figure 2.2   Shows the simplest arrangement of viral coat protein molecules on the surface of a dodecahedron, containing 60 capsomeres that consist of identical protein molecules. Dotted lines indicate groups of five capsomeres that form the vertices of this icosahedron.

In the isometric viruses the protein subunits are usually arranged in groups of equilateral triangles. The simplest and by far the most common shape of virions show **icosahedral symmetry**. An icosahedron is a body with 20 equilateral triangular facets and 12 vertices. It appears that the icosahedral shell represents the most efficient design for biological containers, one that requires minimal energy in assembly and is the simplest form consisting of 60 subunits (3 per facet) and is schematically presented in figure 2.2.

Large icosahedral virions are built from more than 60 subunits, the vertices composed of morphological units or capsomeres containing five subunits (which are termed as pentamers, or pentons, for short) shown in figure 2.3. If these extra subunits are identical to those in the pentamers, as is often the case in large viruses, they cannot be added in random configurations or numbers. In fact, Caspar and Klug realized that stable, *symmetrical icosahedra can be constructed from identical subunits only by using multiples of 60 subunits*. They showed that this can be done by dividing each triangular facet into smaller triangles defined by three subunits as shown clearly in figure 2.3 that subunits from the smaller triangles located on adjoining large triangular facets appear to be arranged in clusters of six (called **hexamers**, or **hexons**). They also introduced the concept of **quasi-equivalence** to describe the observation that identical subunits could form both pentamers and hexamers that, in turn, could form stable icosahedral virions.

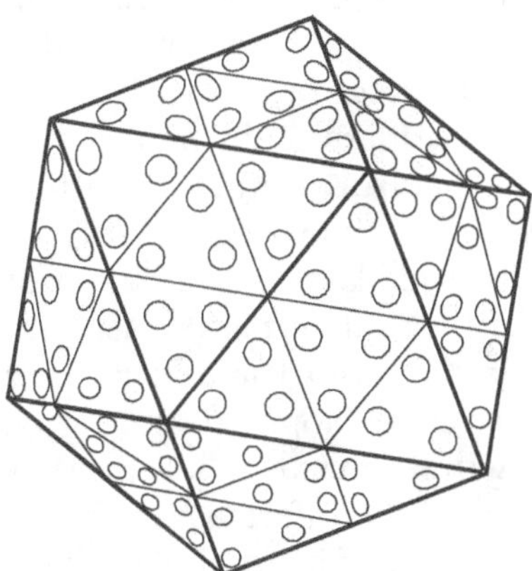

Figure 2.3   Shows arrangement of structural capsomeres in large icosahedra with more capsomeres.

In most plant and bacterial viruses identical coat protein subunits do actually make up both the pentamers and hexamers; but some animal viruses, such as adenoviruses, utilize different protein subunits in their pentons and hexons. The total number of triangular facets needed to form large, geometrically stable bodies can be described mathematically by the so-called **triangulation number**, T, which is 3 (T=3) for most of the unenveloped icosahedral viruses of animals, plants, and bacteria. It is however, 1 (T=1) for Picornaviridae, and 4 (T=4) for the one known member of Tetraviridae, the *Nudaurelia* b virus of insects.

The Virus Infection Cycle

The earliest stages of the virus infection cycle are most generally called **adsorption** and **entry**, during which virions attach to the cell and the viral nucleic acid enters the cells, with or without other virion components, depending on the virus.

During the next stage of infection, the main processes of viral replication occur: production of nucleic acids and proteins. Since no complete infectious particle is found during this stage, even if the cells are disrupted, early virologists called this stage the **eclipse period**—stage of intracellular phage growth during which the infected cells contain no material capable of initiating a plaque. A more general term for this developmental stage is the **latent period**—stage of intracellular phage growth during which there is no increase in phage titer; time that elapses between moment culture is infected with phage to the moment the first cells lyse. The distinction between the two terms is that the eclipse period lasts only until virions are formed inside the cell, whereas the latent period lasts until newly assembled virions are released from the cell. Rise period is the stage of intracellular phage development during which the number of plaque forming units increases; it is the time span during which more and more of the infected bacterial cells lyse. Plateau of infectivity is the stage of intracellular phage development when all the bacteria that are going to lyse have done so. Vegetative phage is the noninfective, intracellular form of phage; it contrasts to mature phage and equals the link between parental and progeny phage. The period of **assembly** of virions from their structural components and nucleic acids occurs late during infection, and the final steps of **maturation** of virions in infectious form may continue even during their release from the infected cell. The release of new virions is commonly called the burst indicating the sudden disruption of the infected cell that occurs during many viral infections. Burst size is the average number of progeny phage particles produced per infected bacterium. This cycle is represented in figure 2.4.

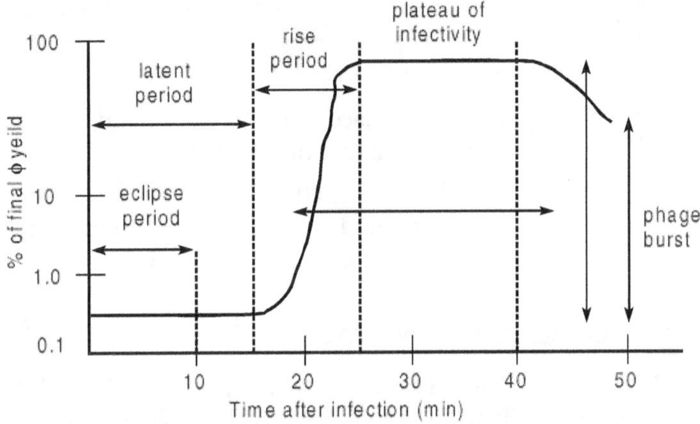

Figure 2.4   Virus infection cycle.

Bacteriophages were jointly discovered by Frederick Twort in England and by Felix d'Herelle at the Pasteur Institute in France. Four types of bacteriophages have been widely used in biochemical and molecular biological research. Most of these infect *Escherichia coli*.

T SERIES OF DNA BACTERIOPHAGES

The T series of bacteriophages had a central role in the development of molecular biology. In 1944, at the instigation of Max Delbruck, the phage group at Cold Spring Harbor agreed to concentrate their research on 7 bacteriophages, all of which were active against *E. coli B*, which had first been isolated by Demerec and Fano. Until this point, different scientists worked with different phages, as a result it was difficult to compare results.

This decision, was greatly criticized since it led to the neglect of some other important aspects of bacteriophage biology, most notably lysogeny. Nevertheless, the study of the T bacteriophage has contributed a great deal to our understanding of molecular biology and genetic regulation.

- The Hershey–Chase experiment used T2 bacteriophage.
- Volkin and Astrachan's experiments which led to the discovery of mRNA were based on the study of T2 bacteriophage infection.
- Seymour Benzer used T4 bacteriophage in his classic fine structure analysis of the gene.
- A T7 bacteriophage promoter has been incorporated into a number of different cloning and expression plasmids such as pTZ.

Hershey–Chase Experiment

In 1951 bacteriophages were cultured such that their DNA was labelled with ^{32}P and their protein was labelled with ^{35}S. These labelled phages were allowed to infect *E. coli* cells. Ten minutes later, the infected bacteria were subjected to shearing forces in a blender followed by centrifugation. The ^{32}P labelled DNA was associated with bacterial cells and that of ^{35}S was found in the supernatant which confirmed that DNA entered the cell.

CLASSIFICATION OF BACTERIOPHAGES

The major criteria for classification were based on chemical composition and morphology. Viral nucleic acid is either DNA or RNA, arranged as single-stranded (ss) or double-stranded (ds), linear or circular form. This is surrounded by protein capsid that can be symmetrical or helical.

PLAQUE ASSAY

Bacteriophages produce plaques in a lawn of host bacteria growing over the surface of an agar plate. Clearance of zones of bacterial lysis was observed against turbid bacterial growth background. The clear spots in the bacterial lawn are called *plaques*, each containing thousands of phages. The plaque assay is a means of titrating the phages. The following table summarizes the properties of the seven T series bacteriophages.

Table 2.1   Characteristics of different phages.

Shape	Nucleic acid	Virus	Characteristics	Number known
Tailed	dsDNA L	T2	Contractile tail	450
		Lambda	Long,non-contractile tail	800
		T7	Short tail	300
Cubic	ssDNA C	φX174	Large, knob-like capsomeres	25
	dsDNA C	PM2	Lipid-containing capsid	2?
	dsDNA L	PRD1	Double coat, lipids, pseudo tail	4?
	ssRNA L	MS2	Icosahedral shell	35
	dsRNA L	φ6	Lipid-containing envelope	1
Filamentous	ssDNA C	Fd	Long rods	17
		MV–L1	Short rods	10
Pleomorphic	dsDNA C	MV–L2	Lipid-containing envelope	3?

L → Linear; C → Circular

The genome of all of the above phages consist of a single linear molecule of dsDNA. However, circular forms and/or circular permutations exist. All undergo lytic growth exclusively. The T-even phages, viz., T2, T4 and T6, are all related serologically and all have large genomes. Seymour Benzer used T4 bacteriophage in his classic fine structure analysis of the gene. It has a large genome 168,895 bp in length. T4 was the first "prokaryotic" organism in which evidence of gene splicing (i.e., introns) was found.

Table 2.2   Properties of T-series phages

Name of the phage	Plaque size	Head (nm)	Tail (nm)	Latent period (min)	Burst size
T1	medium	50	150 ×15	13	180
T2	small	65 ×80	120 ×20	21	120
T3	large	45	invisible	13	300
T4	small	65 ×80	120 ×20	23.5	300
T5	small	100	tiny	40	300
T6	small	65 ×80	120 ×0	25.5	200–300
T7	large	45	invisible	13	300

The electron micrograph image of T4 from the ICTV database is shown in figure 2.5. The T-odd phages fall into three serological groups: T3 and T7 are related to each other but not to T1 or to T5 which are unrelated. The T7 genome was

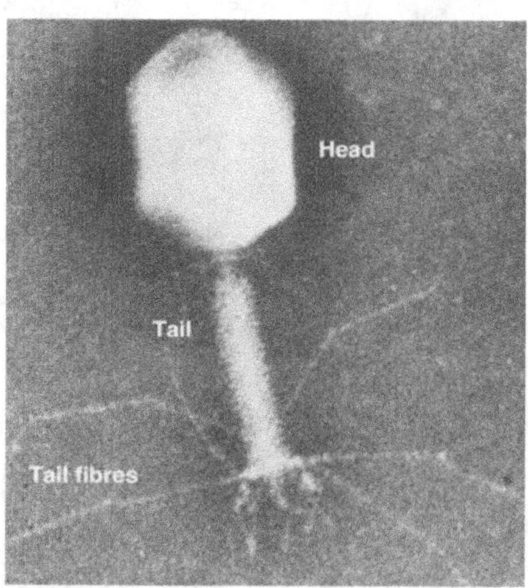

Figure 2.5   Electron micrograph of T4 bacteriophage.

sequenced in 1983; it is 39,937 bp in length. T7 bacteriophage has been used for the study of DNA replication because of its linear chromosome and the problems that it poses for DNA replication and also because it encodes its own DNA polymerase. A modified form of T7 DNA polymerase has been marketed as the very popular sequencing enzyme sequenase. T7 promoters also require a special RNA polymerase; as a result, they have been incorporated into a number of cloning/expression vectors.

Temperate Bacteriophages

Temperate bacteriophages have "alternative" life cycles. After infecting a cell, they can undergo a typical lytic growth cycle or they become dormant in a lysogenic growth cycle. The phage is still present in the cell as a prophage and under certain conditions, such as UV irradiation, it becomes active and resumes a lytic growth cycle.

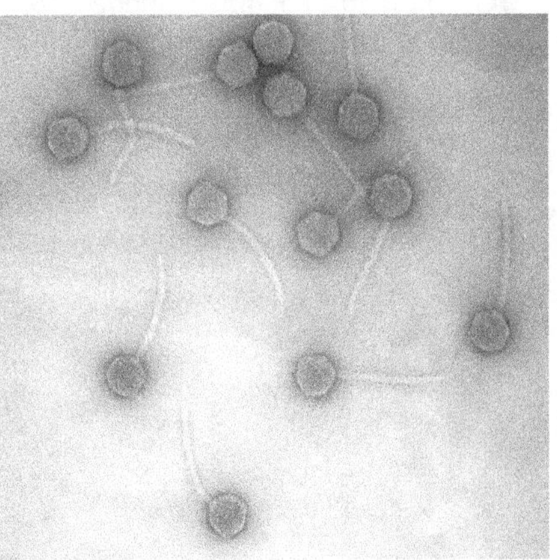

Figure 2.6 Bacteriophage lambda.

Bacteriophage lambda, which infects *E. coli*, is the classic example of a temperate bacteriophage (figure 2.6). It has a linear dsDNA genome, of 48,502 bp, which circularizes after infection (figure 2.7). Bacteriophage lambda has been one of the workhorses of molecular biology particularly as a model system for understanding gene regulation. It has also played an important role as a cloning vector. Characteristics of some important genes of phage λ is stated in Table 2.3.

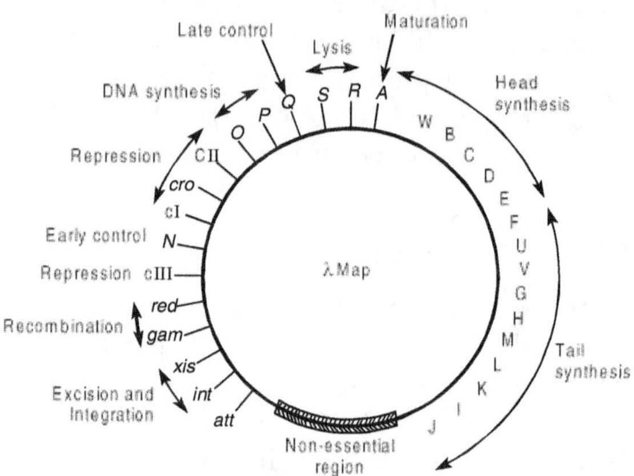

Figure 2.7   Genetic map of lambda phage showing important genes essential for phage life cycle.

Other examples of temperate bacteriophage are P1 (figure 2.8), P22 (which infects *Salmonella*), P4 (11624 bp) (figure 2.9) and Mu (figure 2.10).

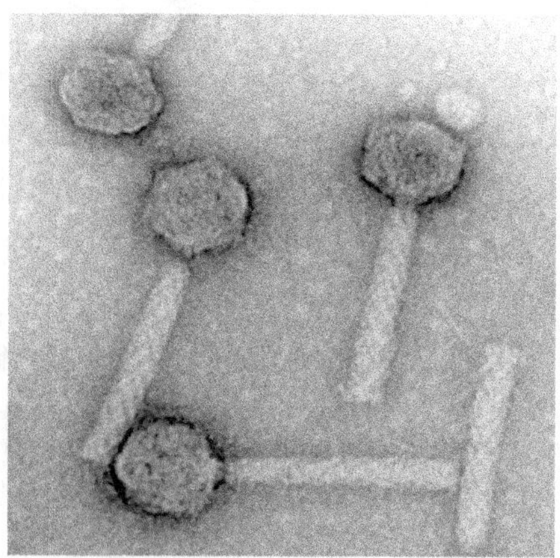

Figure 2.8   Bacteriophage P1.

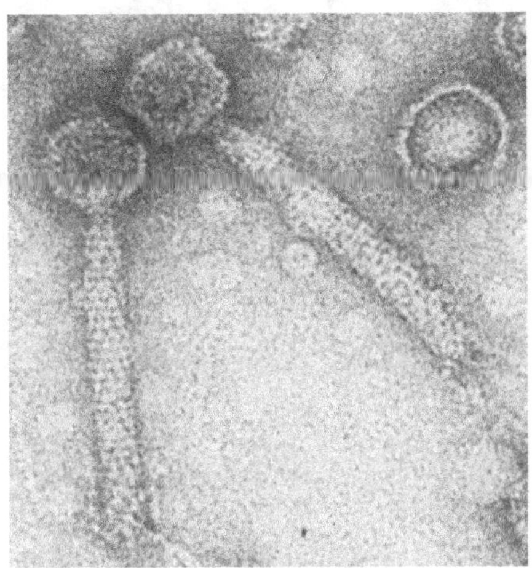

Figure 2.9   Bacteriophage P4.

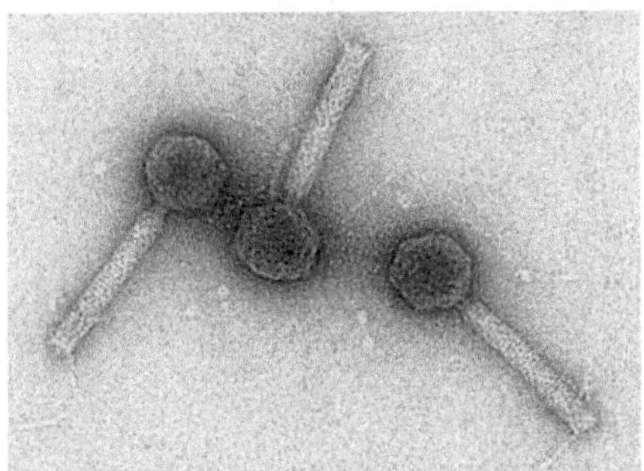

Figure 2.10   Bacteriophage Mu.

Table 2.3   Description of important phage λ genes and their function.

Genes	Function of gene products
A,W,B,C,D,E	Associated with DNA maturation and head proteins
F,Z,U,V,G,H,M,L,K,I,J	Code for phage tail protein
att	Site for integration of phage DNA into host chromosome

(Contd.)

Table 2.3   Contd.

Genes	Function of gene products
int	Integrase protein required for site-specific recombination with chromosome
xis	Excisionase protein forms complex with integrase and functions in excision of prophage
gam	Protein required for rolling circle replication
red	Protein involved in recombination of phage
N	Anti-termination protein acting at (termination sites) t_L', t_R' and tR^2
O,P	Intiation of phage DNA replication
Q	Anti-termination protein acting at t_R'
cIII	Activates *cro*, codes for *cI* and *int* gene products
P_L	Leftward promoter that promotes transcription initiation from N through *int*
P_R	Rightward promoter that initiates transcription from *cro* through O region
cI	Repressor protein–inhibitor of transcription from P_L and P_R
cro	Protein inhibitor of *cI* synthesis

Small DNA Bacteriophages

The small DNA bacteriophages have ssDNA genomes which replicate as dsDNA intermediates. They encode 10–12 proteins. Two groups of these phages can be distinguished.

Spherical phage   The spherical phage (PhiX174, G4, S13) are broadly similar to the filamentous phage. The capsid is icosahedral not helical and is not enveloped (these phages lyse the host cell). Their genome consists of a circular ssDNA molecule. A well-known example is PhiX174 (figure 2.11), which was the first genome to be sequenced by Fred Sanger's group in 1976. Its genome of 5386 bp coded for 11 genes, includes several examples of overlapping genes. The coding frames for 7 proteins overlap: A* is a truncated form of A; B is coded within A in a different reading frame; K is encoded in a third reading frame at the end of A which extends into and overlaps with that of C; E is coded within D in a different reading frame. These were the first examples of overlapping genes. Other relatives of PhiX174 are G4 and S13.

Figure 2.11   Bacteriophage PhiX 174.

Filamentous phage   As described above, the filamentous phage (M13, fd, f1) have a filamentous capsid with a circular ssDNA molecule. The genome, 6407 nt (nucleotide), contains 10 genes but none possess a lysis protein. Virions are enveloped. The filamentous phage will only infect *E.coli* cells carrying the F plasmid since the phage must adsorb the F pilus to gain entry to the cells. Their life cycle involves a dsDNA intermediate replicative form within the cell which is converted to an ssDNA molecule prior to encapsidation. This conversion is the major reason for the great utility of these phages as a molecular biological laboratory tool, providing an easy means to prepare ssDNA for DNA sequencing.

The best-known example is M13 bacteriophage which has been adapted for use as a cloning and sequencing vector. The wild-type M13 genome is 6407 bp in length; the modified cloning vector is 7249 bp in length. Other relatives of M13 are fd and f1.

RNA BACTERIOPHAGES

The ssRNA bacteriophages are the simplest viruses known. Both families (MS2, R17 and f2 form one family; Q beta forms another) have small genomes (3600–4200 nt; MS2 is 3569 nt) that encode four proteins. In the MS2 family (figure 2.12) the proteins are: a capsid coat protein, a replicase, a lysis protein and an attachment protein that is needed for attachment of the phage to a host cell. In the Q beta form they are: a capsid coat protein which also has lysis activity, a replicase, a minor virion protein and an attachment protein. All have linear ssRNA genomes. Because of this, they have served as useful sources of RNA for research in translation and protein synthesis.

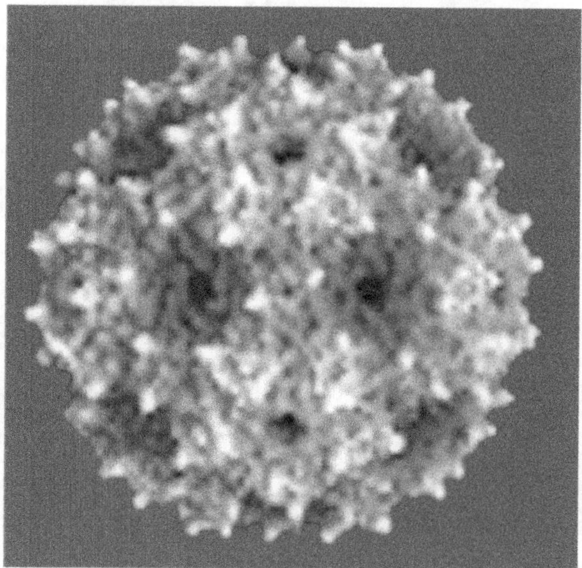

Figure 2.12 MS2 Bacteriophage.

Examples are Q beta, MS2 (3569 nt), R17 and f2. The molecular reconstruction of MS2 bacteriophage, radially depth cued, as solved by X-ray crystallography is shown in figure 2.12.

DNA Bacteriophages

The T bacteriophages, designated T1 through T7, are dsDNA viruses that infect *E. coli*. The T-even bacteriophages (T2, T4 and T6) are composed of a head, tail, tail plate and tail fibres (figure 2.5). The capsid or protein coat of a virus is composed of individual proteins known as capsomeres. The capsid of bacteriophage head surrounds a compartment that contains the nucleic acid, which in these phages is a linear DNA molecule (figure 2.13). This DNA is protected against degradation by environmental nucleases because of its location inside the bacteriophage head.

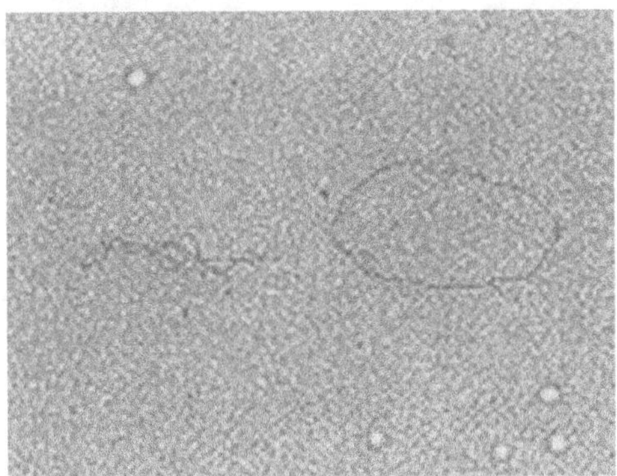

Figure 2.13   Linear DNA molecule of T7 phage.

The tail assembly is involved in attaching the phage to the host cell and injecting the phage DNA into the host's cytoplasm. The phage tail is composed of a core surrounded by a sheath and terminal tail plate. The spikes and tail fibres emanating from the tail plate specifically binds to receptors on the outer membrane of the *E. coli* cell wall. Following adsorption to the host cell, the tail sheath loses proteins in a process that shortens the sheath and propels the tail core through the cell's outer layers (figure 2.14). During this process, the DNA is extruded out of the bacteriophage head and into the cell's cytoplasm. This infection usually kills the host cell as the virus destructs the cell's synthetic activities for its own replication.

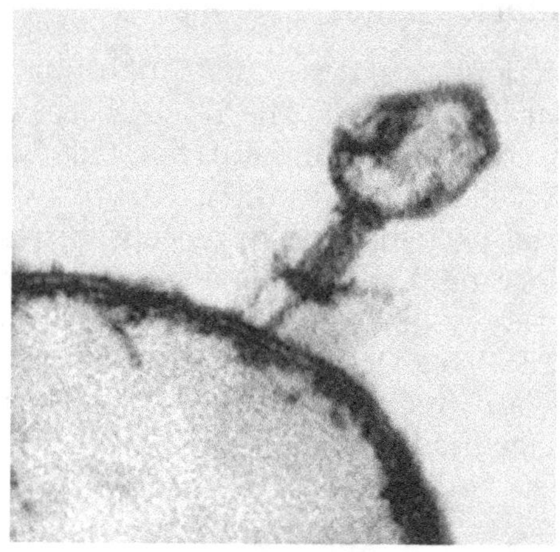

Figure 2.14   Adsorption of a T-phage on the cell membrane.

Bacteriophage Growth Curve

The *lytic cycle* of bacteriophage replication is initiated by the injection of bacteriophage DNA into a host cell and culminates in the release of progeny viruses upon lysis of the host. The stages of the phage replication cycle are portrayed in the *one-step growth* experiment that follows the production of bacteriophages in a culture of host cells, all of which were infected simultaneously at zero time. The experiment depicted in figure 2.15 was initiated when *E. coli* was infected with T1 bacteriophage such that the culture contained 10^5 virus-infected bacteria per millilitre. The number of infected units (infected bacteria plus newly produced T1 bacteriophages) was measured by the plaque assay method and plotted as a function of time.

The results demonstrate that bacteriophage replication occurs in two distinct phases. The *latent period* is the time interval following infection (23 min. during which no increase in the

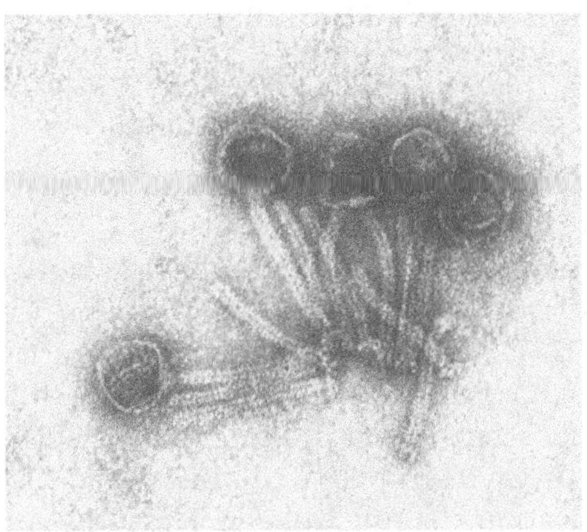

Figure 2.15 T1 Bacteriophage.

number of infection units is observed, which is followed by a *rise period*, approximately 10 min), in which the number of the infective units increases dramatically. In the experiments shown, the increase is a hundred fold. The difference between the initial and the final number of infected units is called the *burst size*.

The "One Step Growth Curve"

In the 1930s, Emory Ellis and Max Delbrück performed a classic experiment which revealed the fundamental nature of virus replication. Although this experiment involves a bacteriophage, the principles apply to *all* viruses, (but not all viruses show such clear-cut or rapid replication cycles, e.g. HIV). The replicative process is called the lytic cycle because each infected bacterium is lysed by viral enzymes produced during phage replication.

Do bacteriophages divide inside the cytoplasm during the latent period? Experiments show that they do not. If infected cells are experimentally lysed during the first 11 min of the lytic cycle, no bacteriophages are recovered from the cell lysate. Following this 11 min. *eclipse period*, the number of infective bacteriophages continually increases inside the cell until the number of viruses reaches a plateau characteristic of the maximum burst size.

Replication of a T-Even Bacteriophage

The replication of T4 bacteriophage in *E. coli* was chosen to demonstrate the principles of viral replication.

Between 1939 and 1941 Milislav Demerec isolated set of phages from *E. coli* and termed it as T1, T2, ... T7. T4 phage has a molecular weight of 110 million, length of 55 mm, head is 60 × 90 nm, DNA (of 173 kbp) contains 5-hydroxy methyl cytosine (HMC) instead of cytosine (C). DNA of T4 is terminally redundant (i.e., sequence of bases repeated at the ends ~ 1600 bases). Characteristic feature of T4 phage DNA is that the DNA molecule differ from phage to phage even in phages obtained from replication of a single phage. Thus DNA undergoes cyclical permutation in the number of bases at termini. The genome map of T4 phage is shown in figure 2.16.

Three important features of T4 phage replication cycle are as follows:

(1) Source of T4 DNA nucleotides—this is obtained by degradation of host DNA into dNMP through enzyme that cuts DNA at C residue provides sufficient material for 30 phages remaining precursors are obtained from de novo pathway.

(2) Synthesis of HMC—two phage enzymes play a key role in conversion in step 1 dCMP is converted to dHMP by T4 hydroxy methylase, dHMP to dHDP by HM kinase. dHDP to dHTP is done by host enzyme nucleoside phosphate kinase dCMP → dHMP → dHDP → dHTP → HMC → Glucosylation of HMC

(3) Prevention of incorporation of dCTP—because dCTP containing DNA is a substrate for phage nucleases. This is achieved by phage enzyme dCTPase that specifically degrades dCDP and dCTP to dCMP which is converted into dump by phage deaminase.

T4 phage life cycle features is stated below:

t = 0   Adsorbs to cell wall, injection of phage DNA within seconds of adsorption t = 1   Synthesis of host DNA, RNA, and protein begins

t = 2   Synthesis of first mRNA of phage begins

t = 3   Degradation of bacterial DNA begins

t = 5   Phage DNA synthesis is initiated

t = 9   Synthesis of late mRNA begins

t = 12 Completed head and tail appears

t = 15 First complete phage particle appear

t = 22 Lysis of bacteria, burst size is 300

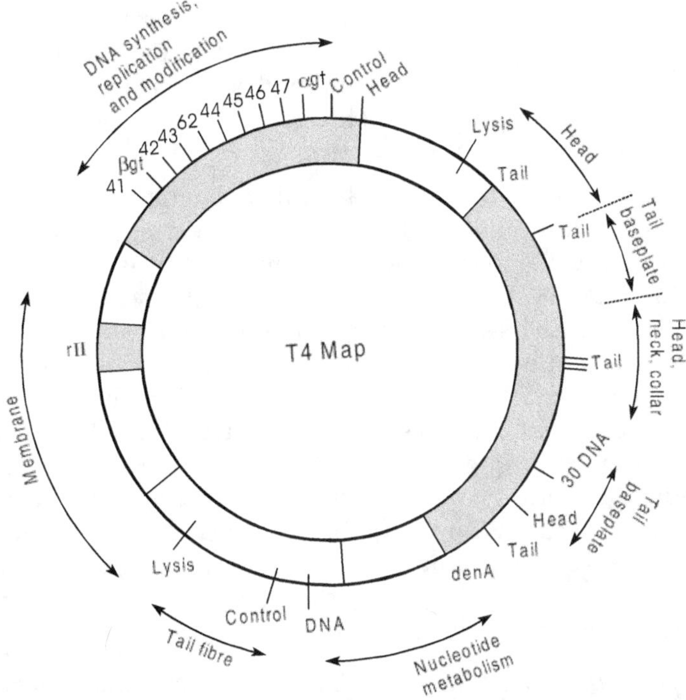

Figure 2.16 Map of T4 DNA showing important genes. 135 genes identified, of which 82 are metabolic and 53 are particle assembly genes. Shaded boxes indicate location of essential genes. "Control" indicates genes needed to initiate transcription.

Adsorption   T4 bacteriophage adsorbs to a specific lipopolysaccharide receptor on the outer membrane of the *E. coli* cell wall. The tail fibres help to position the bacteriophage over the cell's receptor in the initial steps of adsorption. Not all *E. coli* cells possess this lipopolysaccharide; and cells that do not are resistant to T4 infection.

Other bacteriophages bind to different cell surface receptors. Bacteriophage receptors include the pili of *E.coli*, the Vi-antigen of *Salmonella*, glucosylated teichoic acid of *Bacillus subtilis*, and portions of the outer membrane of the gram–negative bacterial cell wall. Because adsorption is a specific biochemical reaction, bacteriophages can infect only certain host cells. This adsorption is so specific that it is used to classify bacteria within a given species. Strains having the same bacteriophage receptors are classified together, just as bacteria possessing the same surface antigens are classified together.

Penetration   Many of the tailed bacteriophages have a contractile mechanism that propels the central tail core through the cell wall and into the bacterium's cytoplasmic membrane. The T4 tail sheath shortens when tail-sheath proteins are released into its surroundings. Tail "contraction" is thought to be powered by ATP associated with the tail rings. Penetration through the cell wall is assisted by an enzyme lysozyme in the tail core, that degrades cell wall peptidoglycan. The T4 DNA then passes through the central tail core into the host cytoplasm, while the empty protein capsid remains attached to the cell surface. More than one phage can adsorb to a host cell and inject its chromosome. If enough bacteriophages are adsorbed, they create so many holes in the cell that it dies, a phenomenon called lysis from without.

Early protein synthesis   The T4 bacteriophage genome is arranged in segments that are transcribed at different times during the replicative cycle. Early protein is made after the host cell's RNA polymerase transcribes a short segment of viral DNA into mRNA. The host's RNA polymerase recognizes the viral promoter that governs the transcription of the genes coding for early protein synthesis. One of the early proteins to be synthesized is the virus own DNA–dependent RNA polymerase, which "recognizes" the other promoter regions on the viral DNA. The mRNAs coded for by the viral DNA are translated into proteins in the host cell's cytoplasm. The host cell's ribosomes, tRNAs, amino acids, ATP, other forms of energy and enzymes are utilized for viral protein synthesis.

Early proteins are involved in commandeering the host's cellular machinery and in manufacturing progeny bacteriophages. Early proteins include nucleases, synthetic enzymes, and the previously mentioned RNA polymerase. The bacteriophage nucleases, and protein kinases specifically prevent the cell from replicating its own DNA. Bacteriophages also make enzymes that are involved in the biosynthesis of precursors used to synthesize bacteriophage DNA. The actual synthesis of phage DNA is done by viral DNA polymerase, formed early in the replication cycle, that makes the DNA used in the final assembly of the virions. Some bacteriophages also make an RNA polymerase that synthesizes the mRNA used in late protein synthesis.

Late Protein Synthesis   Towards the end of the replication cycle, T4 synthesizes *late proteins*, including the structural proteins necessary for virion self assembly, the enzymes involved in maturation, and the proteins used in the release of bacteriophages from the cell. Bacteriophage DNA is synthesized in the cell's cytoplasm at the same time the viral structural proteins are being synthesized on the host's ribosomes. At this stage of the replication cycle, the cell becomes a repository for pools of all the bacteriophage proteins and copies of viral DNA.

Assembly   Bacteriophages mature through a sequential process of self-assembly that is governed in part by the laws of quaternary protein structure. The protein capsomeres join to form bacteriophage capsids at the same time they are filled with bacteriophage DNA. Concomitantly, the tails are formed from their component parts. The DNA–containing capsids combine with the tails before the tail fibres attach , to complete the process of phage assembly (figure 2.17).

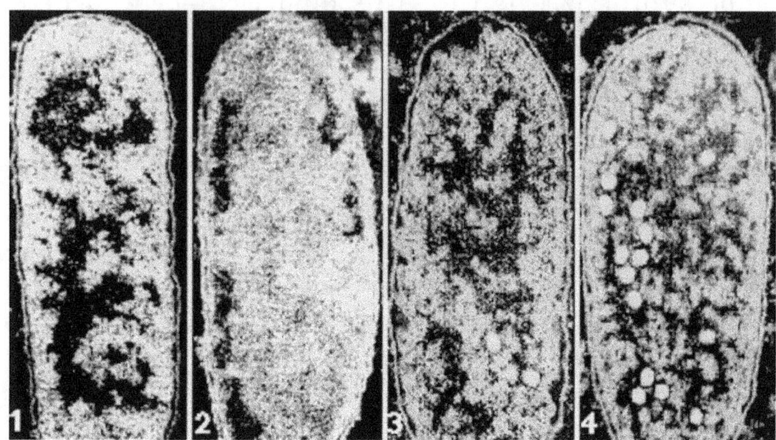

Figure 2.17 Process of phage assembly (1) *E. coli* at the time of infection by phage (2) 5 minutes later, the bacterial DNA has changed appearance (3) 15 minutes later, DNA is gone, replaced by vacuoles containing replicating phages (4) 30 minutes later, many phages are present.

Release   Infective T4 bacteriophages are released into the surrounding environment when their host cells lyse. Release is mediated by a viral protein that damages the cell membrane and lysozyme that damages the cell wall. The combined actions of these enzymes lyse the host cell and release the progeny virions.

Transfer of Bacterial genes by Transduction

On rare occasions, a bacteriophage acquires some of its host's DNA and transfers it to a recipient bacterium by the genetic process known as *transduction*. The recipient cell survives the process because the bacteriophages involved are unable to reproduce. The bacteriophages have no stake in the process—they accidentally acquire bacterial DNA and then act as a delivery system for the donor cell's DNA. Bacteriophages transfer bacterial genes by two basic mechanisms (1) generalized transduction, in which random pieces of bacterial DNA are transferred; and (2) specialized transduction, which involves the process of lysogeny.

Generalized transduction   During the lytic cycle of phage infection, host–cell DNA is broken into small pieces by the viral nucleases synthesized during early protein synthesis. These pieces of host-cell DNA can be incorporated randomly into new phage particles in place of some or all of the phage DNA (figure 2.18). The resulting phage particles are often defective

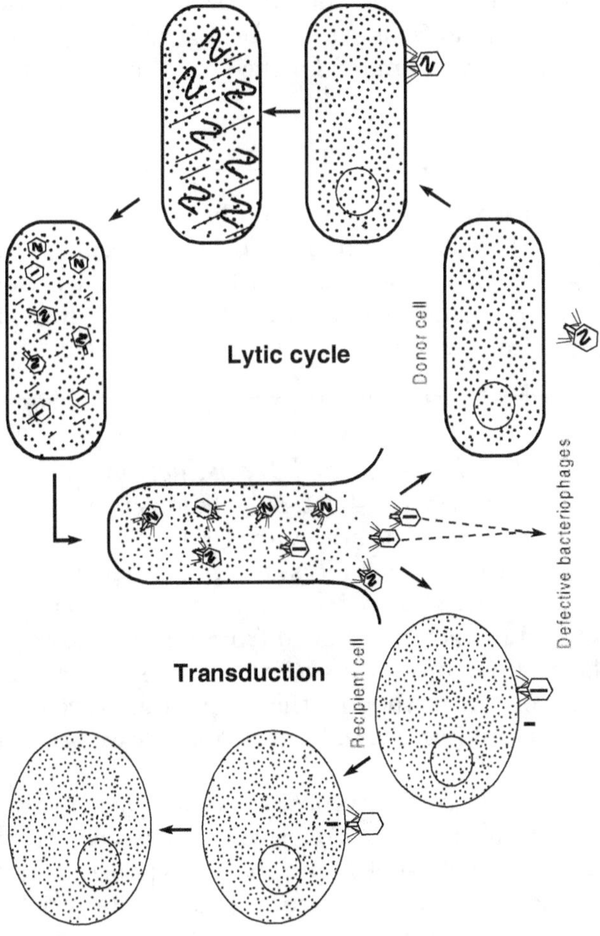

Figure 2.18   Generalized transduction.

because they lack a full complement of their own DNA. When a defective phage infects a bacterial cell, it injects bacterial DNA (from the donor cell) instead of or in addition to phage DNA. Once in the cell, the genes from the donor cell can recombine with the host cell's DNA to become a part of the host–cell genome. This process is known as *generalized transduction* because the phage can pick up any piece of host-cell DNA, so that any gene on the host chromosome can be transferred. Generalized transduction can be used to perform genetic mapping; however, its applicability is limited by the short pieces of DNA that are transferred.

> **KEY POINT**
>
> The volume of the phage head can contain only a limited amount of DNA. When the phage head mistakenly acquires bacterial DNA, it excludes segments of viral DNA required for viral reproduction. Such phages are defective.

Lysogeny and specialized transduction   An unexpected peculiarity of the virus–host–cell relationship was the discovery that many host cells carry viral DNA as part of their genetic material. In many instances, the cell is not destroyed by the presence of these proviruses. This phenomenon was first discovered in bacteria isolated from nature and maintained in the laboratory for months before the presence of the virus was even suspected. Although the bacteria appeared normal, the cultures would occasionally lyse and release bacteriophages.

Any bacteriophage whose genome is an integral part of a host cell's chromosome is referred to as a *prophage* (or provirus). A bacterium that carries a prophage is a *lysogen* and the process of producing a lysogen is termed *lysogeny*. Bacterial strains can carry a prophage through thousands of generations without being adversely affected. The principles of lysogeny are explained below using lambda bacteriophage (λ) as an example.

Temperate bacteriophages are viruses that can exist as both; a replicative form in the cell's cytoplasm, and as a prophage integrated into the host cell's chromosome. Lambda is a temperate DNA bacteriophage that infects *E. coli*. In its replicative form, lambda exists as a circular DNA molecule (physically similar to a plasmid) in the cytoplasm of its host. Its replication results in the destruction of the cell and the release of virulent lambda phage (figure 2.18). Upon infection of another cell, the virulent lambda phages either reproduce via the lytic cycle or become prophages. The process that dictates the outcome of infection by a temperate bacteriophage—lytic cycle or lysogeny—in part depends on the formation of a repressor protein encoded by lambda DNA.

The Lambda Repressor Protein   Lambda DNA codes for a repressor protein that is formed as soon as the DNA enters the cell's cytoplasm. The lambda repressor binds to sites on the phage DNA (figure 2.19) called the right operator. The genes to the right of this operator code for lambda virus production, while the genes to the left code for the lambda repressor protein. When the lambda repressor binds to the right operator, it stimulates the transcription of the repressor gene (left) and prevents the transcription of lambda production genes (right).

Virus production occurs only after the lambda repressor is inactivated. The destruction of the lambda repressor is mediated by the cell's SOS response to deleterious environmental effects. When the cell is exposed to UV light or mutagens, the SOS response is activated and certain proteins are formed. One of these proteins, the product of the *recA* gene, has a protease activity that attacks repressor proteins. The lambda repressor protein is inactivated by the recA protease, permitting transcription of the lambda genes to the right of the operator (figure 2.19) and initiating the lytic cycle of lambda phage reproduction.

Integration of the Lambda Genome   Lambda DNA is double-stranded except for short 12-base sequences at its ends (figure 2.20). These single-stranded tails are complementary, so lambda DNA can form a nicked ring structure. Integration is under the control of the lambda *int* gene, which codes for integrase, and

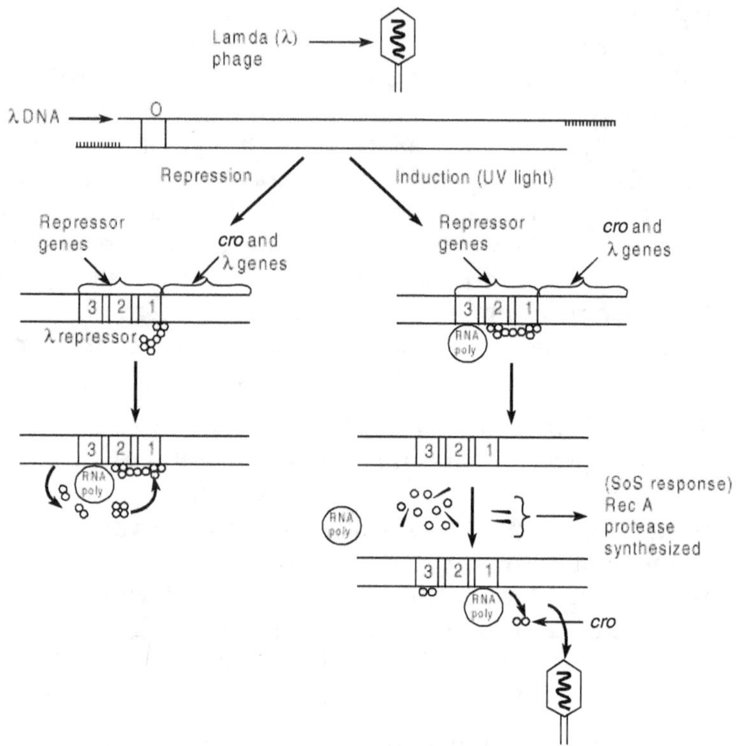

Figure 2.19 Lambda (λ) is a temperate bacteriophage that maintains prophage state by a repressor protein which binds to sites 1&2 on the right operator (O) of λ DNA. This binding activates the genes to the left of the 'O', and λ repressor protein is produced (but not of the genes right of the 'O'). Progeny λ phages are produced when repressor protein is destroyed. Prophage when treated with UV light produces the Rec-A protease that destroys λ repressor protein and thereby activates transcription of *cro* and λ genes. Progeny phages are thus produced.

an integration factor synthesized by the host. Integrase recognizes specific sites on the genomes of both the bacteria and the phage genome called attachment sites (*att*). Because there is only one *att* site on the *E. coli* genome, lambda DNA always integrates at the same site on the bacterial chromosome (figure 2.18). Therefore prophage lambda is always present in the host genome between specific host genes.

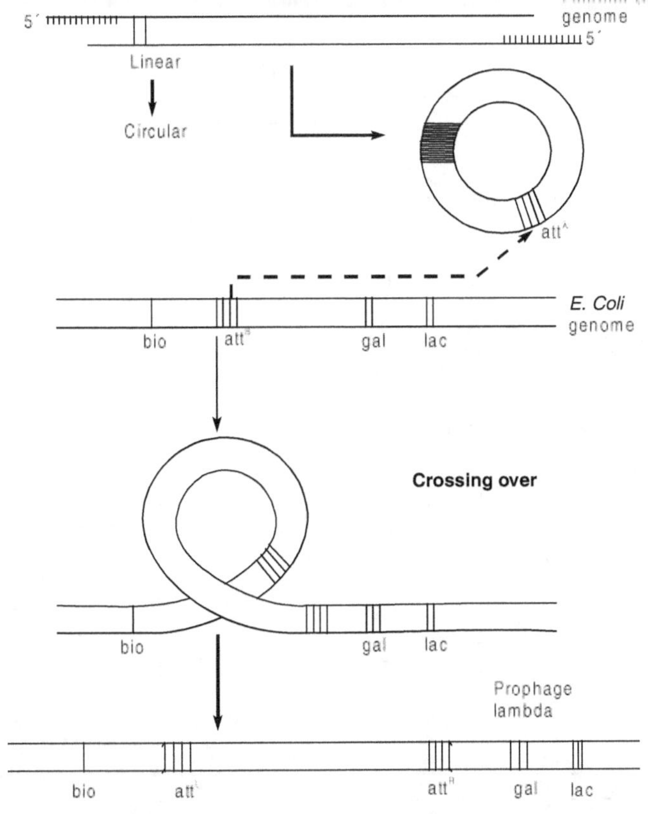

Figure 2.20 Integration of λ bacteriophage into the *Escherichia coli* genome; double-stranded λ DNA spontaneously forms a closed nicked ring prior to being joined with the *E. coli* chromosome by integrase enzyme at the attachment site (att).

Excision of the lambda genome Excision is normally a precise process that results in a replicative form of lambda. Excision is catalysed by the phage enzyme *excisionase* and results in a recombination event to form the circular lambda DNA (figure 2.21). A precise or legitimate recombination occurs when the entire lambda genome is excised. An illegitimate recombination generates an excised circular DNA molecule that is usually defective because essential phage genes have been replaced by short sequences of bacterial DNA.

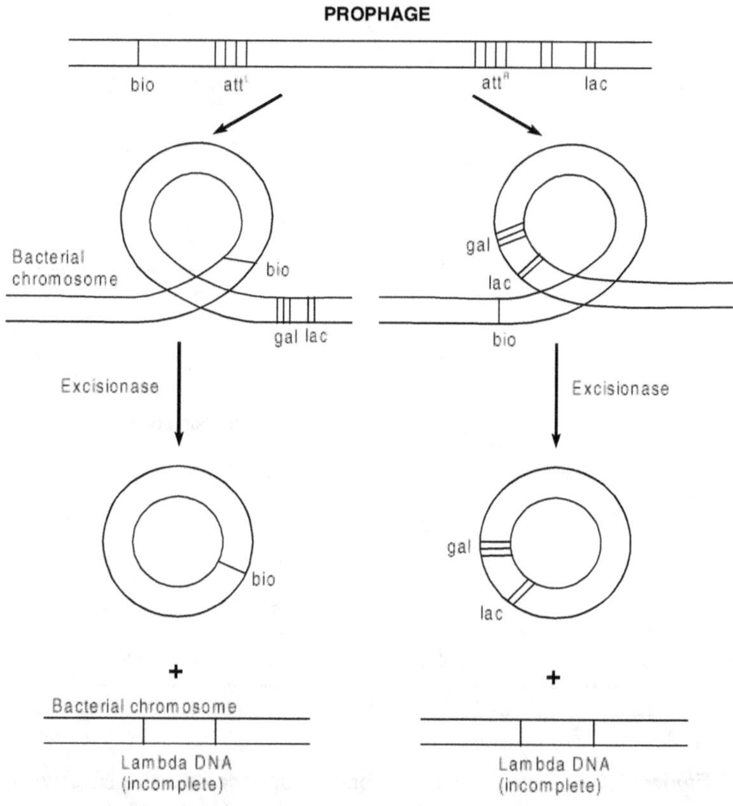

Figure 2.21 λ is a temperate phage that integrates into the host chromosome at a specific site. Error in excision may lead to the formation of incomplete λ genomes that carry host-cell genes (*gal* or *bio*).

Specialized transduction   After excision, the circular lambda genome codes for the replication of progeny phages. A cell with a normal lambda genome goes through a lytic cycle

> **KEY POINT**
>
> Three examples of how DNA segments integrate into the bacterial chromosome are: the F plasmid integrates into the chromosome of an Hfr cell; transposons move between plasmids and the bacterial chromosome; and prophage DNA integrates into the chromosome of a host cell to make it a lysogen.

and produces infectious lambda phages. Cells with defective lambda genomes can produce only defective progeny viruses. Each defective lambda carries bacterial genes that were located adjacent to the attachment site in the bacterial chromosome. When these defective phages infect a recipient *E.coli* cell, it receives both the partial lambda genome and the accompanying bacterial genes. In essence, λ carries bacterial genes to the recipient cell. This type of bacteriophage-mediated genetic transfer is specialized because only the genes located adjacent to attachment sites in the donor bacterial chromosome can be transferred.

PHAGE CONVERSION

Many bacteria are lysogens; that is, they carry viral genetic information in their chromosome as a prophage. These viruses are taking a free ride since they are replicated each time the cell replicates. Normally the prophage's presence goes unnoticed since it does not affect the phenotype of the host cell. Some prophages, however, carry genes that markedly affect the host's phenotype. *Phage conversion* is an alteration of the phenotype of a lysogenized cell by the expression of a prophage gene. The prophage genes of special interest to medical microbiologists are those that code for toxins.

Lysogenized strains of *Corynebacterium diphtheriae* produce diphtheria toxin, which is directly responsible for the symptoms of diphtheria. This protein toxin is coded for by a β phage gene,

> **KEY POINT**
>
> Proviruses are not restricted to bacteria. Humans can carry retroviruses such as the human T cell lymphotropic virus (HTLV–I) and the human immunodeficiency virus (HIV). Other human viruses cause latent infections in which disease symptoms recur months or years after the primary infection. Human diseases attributed to latent viral infection include herpes cold sores, genital herpes, chickenpox and infectious mononucleosis.

so only cultures of *Corynebacterium diphtheriae* carrying the β prophage can cause diphtheria. Similarly, the erythrogenic toxins produced by *Streptococcus pyogenes* and the botulism toxins produced by some strains of *Clostridium botulinum* are coded for by prophage genes. When the lysogenized cells lose their prophage, they also lose the ability to produce the toxin.

REPLICATION OF SINGLE-STRANDED DNA BACTERIOPHAGES

Viruses with single-stranded DNA are either filamentous (F1) or cubic (φX174) viruses. Since they lack a tail and tail fibres, their means of attachment and penetration differs from that of the tailed bacteriophages.

The filamentous bacteriophages attach to the hollow pili extending from the cell wall. The mode of entry into the cell is still unknown: the entire virion could be pulled into the cytoplasm, or their DNA may travel down the pilus to the cytoplasm. φX174 is a cubic bacteriophage that attaches to the lipopolysaccharide of the outer membrane of *E.coli*. After attachment, its single-stranded DNA accompanied by a φX174 surface spike penetrates the cell's cytoplasm.

Phi(φ)X174 phage comprises of circular ssDNA of 5386 nt. The genome encodes three coat proteins H, G and major protein

F. The terminal spike protein (of gene H) enters bacteria along with DNA and plays a major role in entry process. The genetic map and the role of different genes are shown in figure 2.22. Replication of the phage occurs by looped rolling circle method wherein (+) ssDNA form is converted to RFI (+/−) using gene A protein, then (+) ssDNA copies takes place with RFI as template and using gene A protein, the (+) ssDNA copies takes place with RFI as template and using replicase, SSB protein and host polymerase III. The newly synthesized DNA are then packed into phage heads which are later released by lysis.

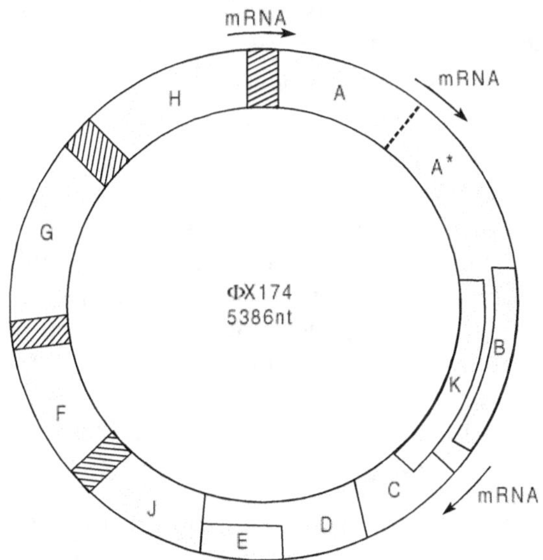

Figure 2.22 Genomic structure of φX174 phage. A involved in replication of phage; B, C, D involved in synthesis and packaging of ssDNA; E in cell lysis; J in internal protein; F major capsid protein; H, G in spike protein. Gene B lies within gene A, gene E lies within D, C overlaps with genes A and D.

The entering single-stranded DNA serves as a template for synthesis of its complementary strand. The cell's DNA polymerase is used since this process occurs before transcription of mRNA occurs. The resulting double–stranded DNA is the replicative strand from which multiple copies of the complementary strand are synthesized.

REPLICATION OF RNA BACTERIOPHAGES

RNA bacteriophages are the simplest of the known viruses. Most RNA bacteriophages are single-stranded, plus RNA coliphages (plus RNA functions as mRNA). They infect *E. coli* after their A protein binds to the cell's F pilus. The A protein and the RNA enter the cell by a mechanism that may involve transfer through the pilus, and once inside the host's cytoplasm, replication commences. About 5000–10,000 progeny bacteriophages are produced in a 30–60-minute replication cycle.

The male coliphages are single–stranded plus RNA viruses (ϕ6 infects *Pseudomonas phaseolicola* and is the only known dsRNA bacteriophage). In the host cytoplasm the plus RNA functions as mRNA for the synthesis of A protein, coat protein and RNA polymerase. This viral RNA polymerase makes a minus RNA strand (replicative strand) using the viral genome as a template. The minus strand is then used to form many plus RNA strands, which combine with coat protein and protein A to form infectious virions. RNA bacteriophages can be extruded from the cell without causing cell lysis.

REPLICATION OF M13 BACTERIOPHAGE

M13 Bacteriophage has a single-stranded DNA genome comprising of 10 genes shown in figure 2.23. After M13 infection, a complementary strand is synthesized, generating a double-stranded replicative form (RF) of the bacteriophage genome. The complementary strand then serves as a template for synthesis of new single-stranded viral DNA by a rolling circle mechanism (except that the second strand is not synthesized until a new cycle of replication is initiated). The single-stranded DNA is cut into genome length fragments and extruded from the cell. This single-stranded DNA is useful in sequencing studies on foreign DNA cloned into M13. In

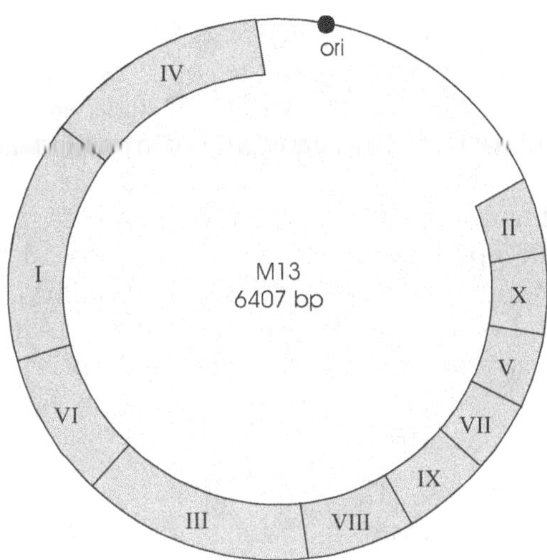

Figure 2.23 Genome of M13 showing the arrangement of genes I to X. Only 507 nt intergenic sequence is free which includes "ori" (origin of replication site).

addition, M13 clones can easily be subjected to site-directed mutagenesis.

When the phage attaches to an *E. coli* cell, this molecule is injected into the cell where most of it is coated with single-strand binding protein (SSB). Since M13 bacteriophage does not code for its own DNA polymerase, it must use the host cell machinery in order to replicate. It is therefore, constrained by the requirements of the host cell replication machinery. Although its ssDNA genome is a perfect template for DNA synthesis, it is not such a suitable template either for RNA synthesis by either RNA polymerase or by primase.

However, although most of the genome is single-stranded, one part of it forms a double-stranded hairpin. This region somehow can serve as a promoter for the host cell RNA polymerase, which transcribes a short RNA primer. Transcription also disrupts the hairpin. DNA Pol III can then take over and synthesize a dsDNA molecule. This dsDNA molecule is known as *RFI-replicative form I*. Further replication of RFI does not proceed by means of theta intermediates but by a type of rolling circle replication. The *gp2 endonuclease*, which is encoded by the phage gene 2, nicks the RFI DNA at a specific site (+ strand origin). *Rolling circle replication* now occurs with displacement of a single strand. Concatemers are not formed; rather the gp2 endonuclease cleaves a second time after one complete copy has been synthesized. Thus the products of this one round of replication are a ssDNA circular molecule (the displaced strand, ligated into a circle) and a dsDNA RFI molecule. The circular ssDNA molecule can now be duplicated by repeating this entire sequence.

In order to synthesize the ssDNA strands that are to be packaged into the capsid, the displaced ssDNA molecules must be coated with a single strand binding protein , *gp5* , which is coded by the phage gene 5. These molecules are then packaged into new phage capsids. The fact that the life cycle of filamentous phage such as M13 includes both a ssDNA phase and a dsDNA phase has been very useful for molecular biologists. Cloning vectors based on M13 have been developed which allow one to clone small DNA fragments and propagate them as phage particles. The dsDNA form permits routine cloning operations. The ssDNA form is ideally suited for the Sanger sequencing protocol and for many protocols for site-directed mutagenesis.

BACTERIOPHAGE REPLICATION— AN UNDERSTANDING WORKOUT

Replication Strategies of Bacteriophage

Studies of DNA replication in bacteriophage have been very valuable because of the insights that have been obtained into replication strategies, mechanisms and enzymology. However, faithful and accurate replication of a genome is easily accomplished only if the genome is circular and is made of double-stranded DNA. If the genome is linear or if it is single-stranded or if it is made of RNA, then special strategies are called for.

Linear Genomes

If the genome is linear then it will progressively shorten with each round of replication unless the organism adopts a strategy for dealing with this problem. The fact that DNA polymerase requires an RNA primer coupled with the fact that DNA polymerases are capable of synthesizing DNA only in the $5'\rightarrow3'$ direction poses a critical problem for the replication of linear DNA molecules. To put in simple words, it is impossible to synthesize two exact copies of a parental molecule under these enzymological constraints. The problem boils down to one of *how do you fill in the ends?*

Consider the figures 2.24 and 2.25.

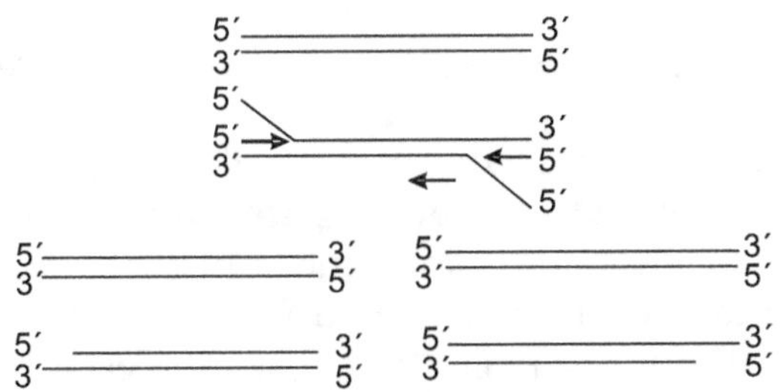

Figure 2.24   Replication of linear DNA molecule.

Assume that replication of the linear molecule is initiated by the synthesis of RNA primers (arrows) at each end. In the above example, these serve as primers for leading strand synthesis which copies each of the two parental strands. The end results are two daughter molecules each with an RNA–DNA hybrid polynucleotide chain.

However, when these RNA primers are removed, we are left with two molecules with single-stranded ends. These ends cannot be repaired or copied by DNA polymerase because of their polarity. Remember that no known DNA polymerase works in a $3' \to 5'$ direction.

Now if we try and follow another round of replication using one of the original two daughters as the new parent, we see the full scope of the problem with replicating linear genomes.

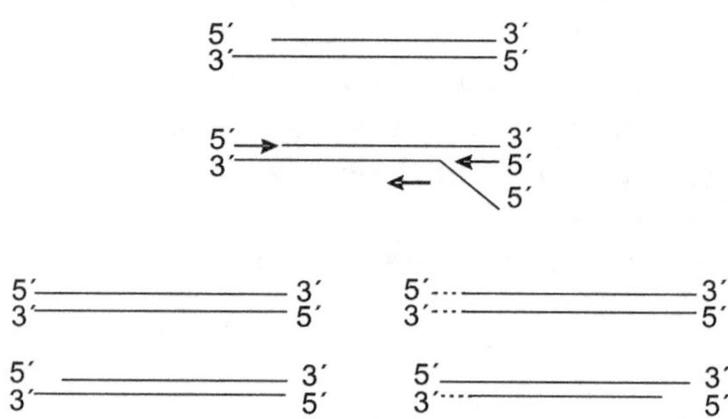

Figure 2.25 Replication of linear DNA molecule.

After another round of replication (b), we recover one molecule that is identical to the parent but the other is not, it is shorter and has lost some genetic material. So the problem with linear genomes is that they will progressively shorten with each round of replication unless some strategy is adopted to prevent this. Two different strategies are described below for overcoming this problem: bacteriophage lambda circularizes its chromosome; T7 bacteriophage forms concatemers.

Other Genomes

If the genome is either an ssDNA or an RNA genome then the organism must use special enzymes or strategies or some combination of the two in order to replicate. An example is discussed in the replication of M13 bacteriophage, which has ssDNA genome.

Rolling circle replication   Replication via theta forms is not the only method by which circular molecules can replicate their genetic information. Another method is *rolling circle replication*, though it generates linear copies of a genome rather than circular copies.

Consider a circular molecule of double-stranded DNA with a nick in one of the two phosphodiester backbones. As long as there is a free 3´ OH end, this can serve as a template for DNA polymerase. When the 3´ OH end is extended, the 5´ end can be displaced in a manner analogous to the *strand displacement* reaction. Synthesis on this strand is also analogous to *leading strand* synthesis. The displaced strand can, in turn, serve as a template for replication as long as a suitable primer is available. Synthesis on this strand is analogous to *lagging strand* synthesis.

If synthesis continues in this manner, the consequence of this mechanism of replication can be the production of *concatemer* copies of the circular molecule. As a result, multiple copies of a genome are produced. A rolling circle mode of replication is seen both during replication of *bacteriophage lambda* where rapid production of many copies of the genome is desired, and in the replication of M13 bacteriophage where only a single copy is produced each time.

Replication of bacteriophage lambda   Bacteriophage lambda contains of a linear dsDNA genome. However, the ends of the genomic DNA are single-stranded and are *cohesive*, i.e., they are complementary to one another. The two cohesive ends, known as *cos* sites, are 12 nt in length. Left 'cos' has the sequence 'GGGCGGCGACCT' and right 'cos' has the complementary sequence 'CCCGCCGCTGGA'.

After adsorption of the phage to the bacterial cell surface and injection into the cell, the chromosome circularizes by means of these complementary cohesive ends. This helps to protect it from degradation by bacterial exonucleases. Circularization is also an essential step if bacteriophage lambda chooses a lysogenic mode of growth. Bacteriophage lambda replicates in two stages.

Early replication Bacteriophage lambda initially replicates by means of *theta* form intermediates. The origin of replication (*ori*) is located within the *O* gene, whose product is required for replication. The gene *P* product is also required for replication.

The *O* gene product has a function analogous to that of DNA *A*. It binds to repeated sequences at the origin and initiates melting of the two strands nearby. The gene *P* product has a function analogous to that of DNA *C*. It helps DNA *B* to bind to the "melted" DNA. Thereafter, the other components of a bacterial replisome can bind and replication ensues. This mode of replication continues for 5–15 min. after replication.

Late replication After 15 min. bacteriophage lambda switches to replication by a rolling circle mechanism. What causes the switch over from one mechanism to the other is not known. As concatemers are synthesized, they must be processed into linear molecules. This occurs by the action of *terminase* which consists of two protein subunits coded by the lambda *A* and *Nu1* genes. Gene *A* codes for a 74 kDa protein; *Nu1* codes for a 21 kDa protein. *Terminase* recognizes the *cos* sites (in its double stranded form) and cleaves them to generate new cohesive ends.

In vivo, processing of the concatemers also requires some of the other capsid proteins and there are length constraints on the amount of DNA that can be packaged. After the first *cos* site has been recognized, the second one must be located within 75–105% of the unit length of the phage chromosome. The ability of the capsid to measure the amount of DNA that is packaged as well as to recognize specific sites is an important factor in the use of bacteriophage lambda as a cloning vector. Bacteriophage lambda derived cloning vectors can only be used to clone DNA fragments that are less than 15 kb in size (the actual size depends on the specific vector).

Replication of T7 Bacteriophage   T7 Bacteriophage has a linear dsDNA genome, 39,937 bp in size. Replication initiates at a site located approximately 5900 bp from the left end of the phage and proceeds bidirectionally. The solution to this replication problem of T7 lies in the left and right ends of the genome. The first 160 bp at the left end are identical with the final 160 bp at the right end. It is this *terminal redundancy* that is the key to its replication. Figure 2.26 shows the products of one round of replication. Although the extent of the single-stranded region is identical to that of the terminal repeat in this figure 2.27, this does not need to be the case.

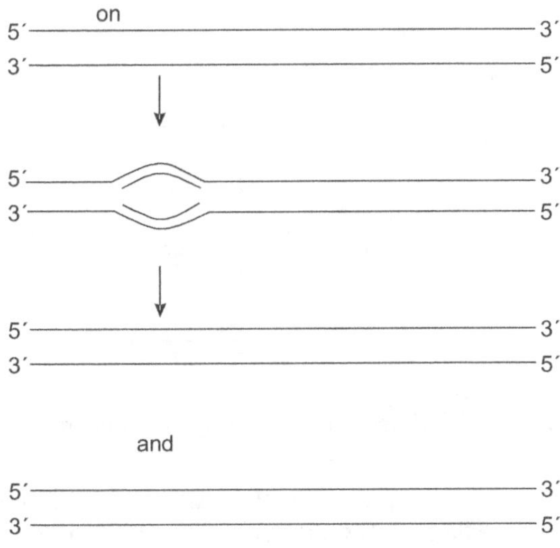

Figure 2.26   Replication of T7 bacteriophage.

The ssDNA at the right end (3′ end after synthesis) of one T7 chromosome is able to anneal with ssDNA at the left end

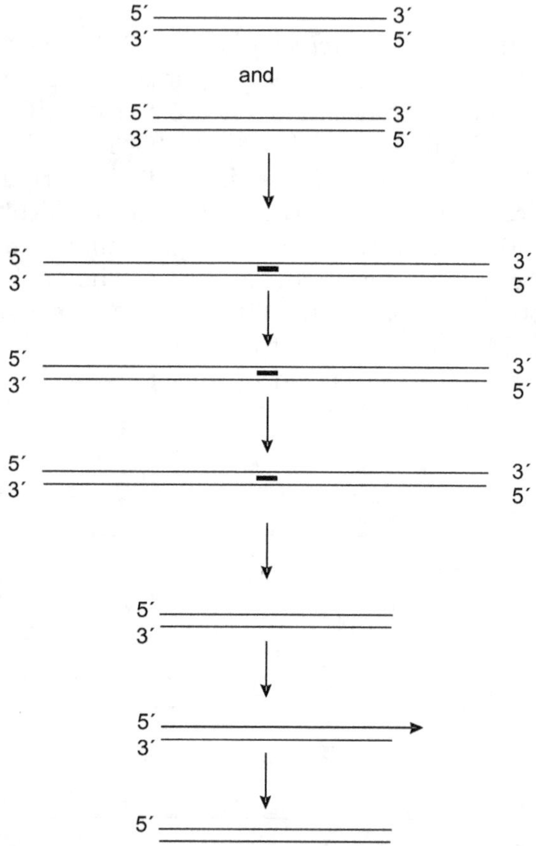

Figure 2.27 T7 phage replication—generation of daughter strands.

(also a 3´ end after synthesis) of another. The remaining gaps can then be filled by *DNA polymerase* and ligated by *DNA ligase*. The resulting dimeric molecule can be cleaved into two again, but this time generating two 5´ overhangs on each daughter, which are now able to act as templates for normal 5´→3´ synthesis: T7 encodes its own DNA ligase (gene 1.3), SSB (gene 2.5) and DNA polymerase (gene 5).

REVIEW QUESTIONS

1. Write short notes on:

    a. Bacteriophage
    b. Hershey–Chase experiment
    c. Plaques
    d. Capsomeres
    e. Burst size
    f. Define lysogeny
    g. Transduction
    h. Temperate phage
    i. Phage conversion
    j. Triangulation number – T
    k. Defective interfering (DI) particles
    l. MOI
    m. Eclipse period
    n. Lytic cycle
    o. Productive vs. nonproductive infection
    p. Cytopathogenic, persistent, latent, transforming, abortive and null infections
    q. Lytic vs. lysogenic
    r. Budding
    s. One-step growth
    t. Latent period
    u. Synthetic (maturation) phase

2. Give an account of the classification of bacteriophages.

3. Write about the replication of a T-even bacteriophage.

4. What is one-step growth curve of a bacteriophage.

5. Give a detailed note on generalized and specialized transduction.

6. Explain the replication mechanism of T7 phage.

7. Not all viruses undergo a lytic cycle. What other outcomes are there for virus infection of cells?

8. What are the steps that occur during a lytic infection process?

9. Give an account of the structural organization, life cycle, DNA replication and phage production of M13.

10. Are there specific differences in challenges faced by infecting bacterial cells and animal viruses? If so explain.

11. What steps occur during a lytic infection process?

12. Not all viruses undergo a lytic cycle. What are the other outcomes of virus infection of cells?

13. How do lytic viruses get released from their host cell?

14. A one-step growth experiment is performed using a bacteriophage and *E. coli*. After infection, some culture fluid is removed and a plaque count is performed. Another sample of culture fluid is treated with chloroform to lyse the bacterial cells before a plaque assay is performed. Draw and label the one-step growth curve for both the intracellular and extracellular viruses.

3 PLANT VIROLOGY

NOMENCLATURE AND CLASSIFICATION OF PLANT VIRUSES

By the early 1930s, three important facts about viruses began to be recognized:

- Viruses can exist as different strains, which may cause different symptoms in the same host plant.

- Different viruses may cause similar symptoms on the same host plant.

- Some diseases may be caused by a mixture of two unrelated viruses.

Several workers have adopted different criteria for their scheme of classification of plant viruses which are as follows:

Johnson J. (1927)	Suggested that a virus should be named by adding the word *virus* and a number to the common name for the host in which it was first found.
Johnson and Hoggan (1935)	Compiled a descriptive key comprising of 5 characteristics, viz. modes of transmission, natural or differential hosts, longevity in vitro, thermal death point and other specific symptoms. More than 50 viruses were identified and placed in groups.
Smith, K. M. (1937)	Viruses were named and grouped according to the generic name of the host in which they were first found. Successive members in a group were given a number. For instance, TMV (tobacco mosaic virus) was *Nicotiana* virus 1, and there were 15 viruses in the *Nicotiana* group.

Holmes (1939)	Based primarily on host reactions and methods of virus transmission. A Latin binomial–trinomial system of naming was followed. For example, TMV has become *Marmor tabaci* (*Marmor* meaning marble in Latin). The classification was based on diseases rather than the viruses, and 53 of the 89 plant viruses considered by Holmes come under the genus *Marmor*.
ICTV	In 1966, the first meeting of the International Committee for the Nomenclature of Viruses was held at Moscow. Subsequent developments in the organization have laid down seven reports on the taxonomy of viruses; this committee is now known as the International Committee for Taxonomy of Viruses (ICTV). A total of 977 plant viruses were listed in its seventh report.

Species Concept of Plant Viruses

In 1991, ICTV accepted the concept that viruses exist as species, adopting the following definition that has formed the basis of modern virus classification.

"A viral species is a polythetic class of viruses that constitutes a replicating lineage and occupies a particular ecological niche."

This enables viruses to be differentiated as species and tentative species, forming the basis of classification system and can be grouped into other taxa on several criteria, the details of which are listed as follows:

I Order	Common properties between several families including: • Biochemical composition • Virus replication strategy • Particle structure (to some extent) • General genome organization

II	Family	Common properties between several genera including:
		• Biochemical composition
		• Virus replication strategy
		• Nature of particle structure
		• Genome organization
III	Genus	Common properties with a genus including:
		• Virus replication strategy
		• Genome size, organization and/or number of segments
		• Sequence homologies
		• Vector transmission
IV	Species	Common properties with a species including:
		• Genome rearrangement
		• Sequence homologies
		• Serological relationships
		• Vector transmission
		• Host range
		• Pathogenicity
		• Tissue tropism
		• Geographical distribution

Viruses are now recognized as "species" or "tentative species" which are viruses that have not yet been sufficiently characterized to ensure that they are distinct and not strains of an existing virus or do not have the full characteristics of the genus to which they have been assigned. Of the 977 plant viruses listed in the ICTV 7th report, 701 are true species and 276 are tentative species.

Acronyms or Abbreviations of Viruses Used

Designation of abbreviation is based on the following principles:

• Abbreviations should be as simple as possible.

- It must not duplicate any other name previously coined and still in current usage.
- The word 'virus' in a name is abbreviated as 'V'.
- The word 'viroid' in a name is abbreviated as 'Vd'.

Use of virus names   The last word of a species is "virus". Suffix (ending) for a

> genus is "…virus"
>
> subfamily is "…virinae"
>
> family is "…viridae"
>
> order is "…virales"

Salient Features of Plant Virus Families

Family Caulimoviridae   This family contains all the plant viruses that replicate by reverse transcription. The genome has circular dsDNA with gaps or discontinuities at specific sites that represent priming sites for DNA synthesis during replication. The primary transcript is asymmetric that has more-than-genome length sequence which is both the template for reverse transcription and the mRNA for some of the genes. There are six genera in this family comprising two groups viz. the caulimoviruses that have isometric particles and the badnaviruses that have bacilliform particles. The genera are distinguished on their genome organization.

Genus	Type species	General properties
Caulimovirus	Cauliflower mosaic virus	Isometric, ~50 nm diameter, genome has 6 ORFs, the 5´ ones being expressed from the more-than-genome-length RNA and the 3´ ORF 6 from a separate mRNA. Mostly transmitted by aphids in the semipersistent manner requiring a virus-coded helper protein.
Soybean chlorotic mottle virus-like	Soybean chlorotic mottle virus	Resembles genus *Caulimovirus* except for differences in genome organization.

Genus	Type species	General properties
Cassava vein mosaic virus-like	Cassava vein mosaic virus	Resembles genus *Caulimovirus* except for differences in genome organization.
Petunia vein clearing virus-like	Petunia vein clearing virus	Resembles genus *Caulimovirus* except for differences in genome organization. Presence of only one large ORF comprising ORFs 1 and 2.
Badnavirus	Commelina yellow mottle virus	Bacilliform, 130 × 30 nm, 3 ORFs present, ORF III expresses a polyprotein processed to give several products. Transmitted by mealybugs in a semi-persistent manner.
Rice tungro bacilliform virus-like	Rice tungro bacilliform virus (RTSV)	Leafhopper-mediated transmission. Genome has 4 ORFs, the 3′ one being translated from an mRNA spliced from the more-than-genome-length RNA.

Family Geminiviridae Viruses belonging to this family possess circular ssDNA genomes. The most distinguishing feature is that these genomes are contained in geminate virus particles consisting of two incomplete icosahedra. It comprises four genera that differ in their genome organization and vectors.

Genus	Type species	General properties
Mastrevirus	Maize streak virus	Have narrow host range, except TYDV and BeYDV which infect dicotyledons, host range limited to species in the Poaceae family. Transmitted by leafhoppers in a circulative, non-propagative manner. Genome is monopartite with single component of ssDNA.
Curtovirus	Beet curly top virus	Important virus of sugar beet, has wide host range. Transmitted by leaf hoppers in a circulative, non-propagative manner.

Genus	Type species	General properties
Begomovirus	Bean golden mosaic virus	Largest of this family have narrow host ranges among dicotyledonous species. Transmitted by white flies. Most members have their genome divided between two DNA molecules.
Topocuvirus	Tomato pseudocurly top virus	Similar genome of Curtoviruses but transmitted by the treehopper *Micrutalis malleifera*.

Family Circoviridae   This family contains two genera, one contains those viruses that infect animals and the second that infects plants. Viruses are small with icosahedral symmetry and 17–22 nm diameter and contain small circular ssDNA genome. Genus Nanovirus (type species—Subterranean clover stunt virus) contains the economically important virus BBTV. The genome has at least 6 circular ssDNA molecules of ~1 kb size encircled in an icosahedral capsid, measuring ~20 nm in diameter.

Family Reoviridae   This family comprises viruses that infect vertebrates, invertebrates and plants. Those that infect plants also infect their invertebrate vectors. Virus particles are complex, made up of one, two or three distinct shells each with icosahedral symmetry and have surface spikes. Genomes comprise 10, 11 or 12 segments of linear dsRNA depending upon

Genus	Type species	General properties
Fijivirus	Fiji disease virus	Have 10 segments of dsRNA encapsidated in double-shelled particles, measure 65–70 nm in diameter. Cause hypertrophy of the phloem leading to vein swelling and enations. Transmitted by delphacid plant hoppers in a circulative propagative manner and infect monocotyledonous plants of the Graminae and Liliaceae families.

Genus	Type species	General properties
Oryzavirus	Rice ragged stunt virus	Similar to Fijiviruses, but distinguished by particle size which measures 75–80 nm in diameter, and by nucleic acid and serological properties.
Phytoreovirus	Wound tumour virus	Have 12dsRNA, segmented genome encapsidated in a double-shelled particle, measure ~70 nm in diameter. They infect both dicotyledonous and monocotyledonous species. Transmitted by leaf hoppers in a circulative propagative manner.

the genus. There are 3 genera that infect plants and are distinguished based on their structure and the number of genomic segments.

Family Partitiviridae   The members of this family have small isometric non-enveloped particles, 30–40 nm in diameter, with a dsRNA two-segmented genome. The smaller segment encodes coat protein and the larger encodes RNA-dependent RNA polymerase. The members contain four genera, two of them infecting fungi. The plant-infecting viruses are known as 'Cryptic viruses' as they cause no, or very few symptoms. They are not transmitted by grafts and have no biological vector, but are highly seed-borne.

Genus	Type species	General properties
Alphacryptovirus	White clover cryptic virus 1	Isometric particles measure 30 nm in diameter, lacking fine structural detail. The genome segments are 1.7 and 2 kbp in size. The capsid is made up of a single protein component of 55 kDa.
Betacryptovirus	White clover cryptic virus 2	Isometric particles measure 38 nm in diameter, show prominent subunits, contain genome segments of 2.1 and 2.25 kbp in size.

Family Rhabdoviridae   The members of this family have characteristic bullet shaped or bacilliform membrane-enveloped particles. The outer surface has glycoprotein spikes. The inner surface contains a layer of matrix protein enclosing the nucleocapsid. The nucleocapsid contains (–) strand RNA and nucleoprotein. There are two genera that infect plants. They are transmitted by insects in a circulative, propagative manner.

Genus	Type species	General properties
Cytorhabdovirus	Lettuce necrotic yellow virus	Members of this genus replicate in perinuclear space or in endoplasmic reticulum and mature by budding into the cytoplasm. Particles are about 60 nm in diameter.
Nucleorhabdovirus	Potato yellow dwarf virus	Viruses of this genus replicate in the nucleus and mature by budding through the inner nuclear membrane into the perinuclear space and are about 90 nm in diameter.

The most common vectors are leaf hoppers, plant hoppers and aphids. Mite-and lacebug-transmitted viruses have also been identified.

Family Bunyaviridae   A large family of viruses, most of which infect both vertebrates and invertebrates. One genus of viruses named *Tospovirus* (type species—tomato spotted wilt virus, TSWV) infect plants and also its invertebrate vectors. Virions of *Tospovirus* genus are pleomorphic or spherical, enveloped and have surface glycoprotein spikes. The genome comprises of 3 ssRNA species, 2 of which have an ambisense arrangement, the largest being of (–)-sense. One of the gene product is involved in cell-to-cell spread of the virus which distinguishes members of this genus from other Bunyaviruses. They are transmitted by thrips in a circulative, propagative manner. TSWV has a host range of more than 925 species belonging to 70 plant families.

Family Bromoviridae   Members of this genus have isometric particles, 26–35 nm in diameter or bacilliform particles whose symmetry is based upon the icosahedron. The genomes have linear (+)-sense ssRNA that are divided between three molecules. The subgenomic RNA for coat protein is often encapsidated. The family comprises five genera.

Genus	Type species	General properties
Bromovirus	Brome mosaic virus	The virions are isometric, ~27 nm in diameter, and are stabilized by a pH-dependent protein–protein interaction. pH > 7.0 make the particles swell and become salt labile. Capsid com-posed of 180 copies of single protein of 20 kDa. Have narrow host range, and transmitted by beetles and some by mechanical means.
Cucumovirus	Cucumber mosaic virus (CMV)	Have isometric particles, ~30 nm in diameter, stabilized by protein–RNA interaction and thus are salt-labile. The capsid comprises 180 copies of single protein of 24 kDa. Have wide host range infecting over 1000 species covering more than 85 plant families. Transmitted by aphids in a non-persistent manner. CMV is divided into 2 subgroups based on serology and nucleotide sequence identity of three RNA molecules.
Alfamovirus	Alfalfa mosaic virus (AMV)	The virions are bacilliform, 18 nm in diameter, and 30–57 nm in length depending on the RNA species. Capsid comprises 24 kDa protein. They are stabilized by protein–RNA interaction and thus are salt-labile. The presence of coat protein or the subgenomic RNA encoding it is required for virus replication. AMV has wide host range and is transmitted by aphids in a non-persistent manner.

Genus	Type species	General properties
Ilarvirus	Tobacco streak virus	They have quasi-isometric particles from being spherical to bacilliform, stabilized by protein–RNA interaction and thus are salt-labile, genome characters are similar to AMV. Mainly infect woody plants and are transmitted by pollen and seeds. Viruses of this genus are divided into 7 subgroups based on serological relationships.
Oleavirus	Olive latent virus 2 (OLV-2)	Have different shapes and sizes ranging from quasi-spherical with a diameter of 26 nm to bacilliform of 18 nm dia., and a length of 55–37 nm. The coat protein is 20 kDa. No natural vector is known.

Family Comoviridae It has three genera. Virions of this family are isometric with T = 1 icosahedral symmetry. The

Genus	Type species	General properties
Comovirus	Cowpea mosaic virus	The capsids are constructed from 2 polypeptides of 40–45 kDa and 21–27 kDa expressed from the smaller RNA species. Transmitted by Chrysomelid beetles.
Fabavirus	Broad bean wilt virus	Capsid construction is similar to comoviruses and are transmitted by aphids in a non-persistent manner.
Nepovirus	Tobacco ringspot virus (TRSV)	Capsids of TRSV are composed of a single polypeptide of 52–60 kDa. Other Nepoviruses may be made up of two or three smaller proteins. Genome organization is similar to comoviruses. Transmitted by Longidorid nematodes, some by pollen, one (BRAV) by mites and few others have no known vector. Have 3 clusters of species, 'a', 'b' and 'c' based on the length and packaging of RNA-2, and on serology.

capsids are made up of one or two coat protein species. The genome comprises 2 (+)-sense ssRNA . The RNAs are expressed as polyproteins that are processed to give functional proteins.

Family Potyviridae   Members of this family have flexuous particles 650–900 nm long and about 11–15 nm in diameter. The genome is (+)-sense ssRNA with a VPg at the 5´ end and a 3´ poly(A) tract. The genome has 2 segments and is expressed as polyproteins that are cleaved into functional proteins. They form circular inclusion bodies in infected cells and are transmitted by a variety of vectors. Have 6 genera and are stated as follows.

Genus	Type species	General properties
Potyvirus	Potato virus-Y (PVY)	Particles are 680–900 nm long and 11–13 nm in diameter. Genome size is of 9.7 kb with multiple copies of single protein capsid of 30–47 kDa. Transmitted by aphids in a non-persistent manner using a helper component. Some of the members are seed-transmitted. Largest of the genera of plant viruses that has 91 species and 88 tentative species. Contain economically important viruses such as PVY, BYMV, PPV and PRSV.
Ipomovirus	Sweet potato mild mottle virus	Particles are 800–950 nm long, genome of 10.8 kb encapsidated in multiple copies of a single protein of ~38 kDa. The natural vector is white fly *Bemisia tabaci* and transmitted in a non-persistent manner.
Macluravirus	Maclura mosaic virus	Particles are 650–675 nm long, genome of 8 kb RNA encapsidated in multiple copies of a single protein of 33–34 kDa. Transmitted by aphids in a non-persistent manner.

(Contd.)

Genus	Type species	General properties
Rymovirus	Ryegrass mosaic virus	Particles are 690–720 nm long and 11–15 nm in diameter, genome of 9–10 kb RNA encapsidated in multiple copies of a single protein of 29 kDa. Transmitted by eriophyid mites.
Tritimovirus	Wheat streak mosaic virus	Particles are 690–700 nm long, genome of 8.5–9.6 kb RNA encapsidated in multiple copies of a single protein of 32 kDa. Host range restricted to Graminae and are transmitted by eriophyid mites in a persistent manner.
Bymovirus	Barley yellow mosaic virus	The virions are flexuous rods 13 nm wide and has 2 lengths, 250–300 and 500–600 nm. Genome is divided between two RNA species, the longer one is 7.5–8 kb in size and the shorter one is 3.5–4 kb in size. Capsid is made up of multiple copies of a single protein of 28–33 kDa. Host range restricted to Graminae, and are transmitted by the plasmodiophora fungus *Polymyxa graminis*.

Family Tombusviridae   Viruses in this family have isometric particles of T = 3 icosahedron, 32–35 nm in diameter. Contain genome of (+)-sense ssRNA with a size range of 3.7–4.7 kb. Each member has a highly conserved RNA-dependent RNA polymerase enzyme interrupted by an in-frame termination codon, which is periodically suppressed. Individual members have relatively narrow host ranges and can infect either monocotyledonous or dicotyledonous plants. Relatively stable in natural environments such as surface waters and soils. It has eight genera that are briefed below.

Genus	Type species	General properties
Tombusvirus	Tomato bushy stunt virus	Genome of 4.7 kb, particle of 32–35 nm, capsid composed of 180 subunits of single protein of 41 kDa, transmission is soil-borne with no obvious biological vector except CNV which is transmitted by fungus *Olpidium bornovanus*.
Aureusvirus	Pothos latent virus (PoLV)	Single species in this genus, has genome of 4.4 kb, particle of 30 nm size, capsid comprising single protein of 40 kDa. Natural trans-mission is through soil or circula-ting solution in hydro-ponics.
Avenavirus	Oat chlorotic stunt virus (OCSV)	Single species in this genus, has capsid protein of 48.2 kDa, particle size of 35 nm, genome length of 4114 nt. Only found in oats (*Avena sativa*). Natural transmis-sion is soil-borne possibly by zoosporic fungi.
Carmovirus	Carnation mottle virus	Isometric particles of 32–35 nm, and 3.88–4.45 kb RNA genome, capsid protein of 36–41 kDa.
Machlomovirus	Maize chlorotic mottle virus (MCMV)	Monospecific genus, particles of 30 nm diameter, genome of 4.4 kb, capsid protein of 25 kDa. Host range restrict to members of Graminae. Transmission is seed-borne, and some have been reported to be by chrysomelid beetles and thrips.
Necrovirus	Tobacco necrosis virus -A	Particles 28 nm in diameter, capsid protein of 30 kDa, genome cons-ists of linear ssRNA of 3.7 kb. Host range include both mono- and di-cotyledonous plants. Infection is restricted to roots. Transmitted by chytrid fungus *Olpidium brassicae*.

(Contd.)

Genus	Type species	General properties
Panicovirus	Panicum mosaic virus	Particles of ~30 nm diameter, genome of 4.3 kb RNA, capsid made up of 26 kDa protein. Has RNA polymerase similar to that of Tombusviridae. Host range restricted to Graminae. Transmission is most likely by mechanical means.
Dianthovirus	Carnation ringspot virus	Genomes are divided between two ssRNA molecules contained in a 32–35 nm particle of icosahedral symmetry. Capsid comprises 37–38 kDa protein. Host range restricted to dicotyledonous plants. Transmission through soil, no biological vector known.

Family *Sequiviridae* Particles of this family are isometric, 30 nm in diameter, contain (+)-sense ssRNA of 9–12 kb in size. The virion is made up of three proteins of 32, 26 and 23 kDa present in equimolar quantities. RNA encodes a polyprotein that is cleaved to give the functional proteins. It has two genera.

Genus	Type species	General properties
Sequivirus	Parsnip yellow fleck virus (PYFV)	The RNA is ~ 10kb and polyadenylated. Transmitted by aphids in a semi-persistent manner and depends on a helper waikavirus AYV.
Waikavirus	Rice tungro spherical virus	RNA is >11kb and is polyadenylated. Transmitted either by leafhoppers or aphids in a semi-persistent non-circulative manner, involving one or more helper viral proteins.

Family *Closteroviridae* Viruses of this family have very flexuous filamentous particles about 12 nm in diameter. Genome is linear (+)-sense ssRNA that may be mono- or bipartite depending on the genus. Host ranges are narrow, usually phloem limited and cause yellowing-type symptoms or pitting or grooving of woody stems. Transmission may be through aphids, whitefly or mealybug depending on genera. This family has two genera.

Genus	Type species	General properties
Closterovirus	Beet yellow virus (BYV)	Flexuous virions 1250–2000 nm long with one end coated with an anomalous coat protein giving a 'rattle snake' structure. Genome contain 15.5–19.3 kb RNA. Natural transmission is by aphids, mealybugs or whiteflies (*Trialeuroides*), dependent of species.
Crinivirus	Lettuce infectious yellow virus	Particles have two modal lengths, 700–900 nm and 650–850 nm, the major coat protein is 28–33 kDa. Genome is bipartite and are separately encapsidated. Natural vectors are white-flies (*Bemisia* and *Trialeuroides*), semi-persistent manner of transmission.

Family *Luteoviridae* Particles are isometric with T = 3 icosahedron, has 25–30 nm diameter, capsid comprises 21–23 kDa protein. Genome has (+)-sense ssRNA of 5.7–5.9 kb. The family has three genera distinguished on genome organization. Phloem limited, cause 'yellow-type' diseases, hence the name luteovirus (In Latin *Luteus* meaning yellow). Transmission is by aphids, in a specific, circulative, non-propagative manner.

Genus	Type species	General properties
Luteovirus	Barley yellow virus (PAV)	Host range restricted to Graminae, cause serious disease.
Polerovirus	Potato leaf roll virus	Infect di- and monocotyledonous plants, cause serious diseases.

Genus	Type species	General properties
Enamovirus	Pea enation mosaic virus-1(PEMV-1)	Caused by a set of two viruses, PEMV-1 (28 nm diameter, icosahedral symmetry) which is a member of Luteoviridae and PEMV-2 (25 nm diameter, quasi-icosahedral symmetry) which is an umbravirus. PEMV-1 infect protoplasts but spread only with the presence of PEMV-2. Transmission is by both mechanical means and by aphids.

Floating Genera

There are currently 20 genera of (+)-sense ssRNA plant viruses that have been placed in families (Table 3.1).

Viruses of Algae

Two groups of algal viruses have been observed based on size. Larger ones are placed under the family *Phycodnaviridae* and smaller algal viruses.

Large algal viruses   Termed as virus-like particles (VLPs), polygonal with varying diameter ranging from 22 to 390 nm size found in many eukaryotic algal species belonging to the Chlorophyceae, Rhodophyceae and the Phaeophyceae families. The most studied viruses are those infecting Chlorella-like green algae. A plaque assay was developed for *Paramecium bursaria* chlorella virus-1 (PBCV-1) during 1983, which infects a culturable Chlorella-like algae. This has led to much knowledge of these virus properties. PBCV-1s are large icosahedra with multilaminate shells surrounding an electron dense core. The outer capsid comprises 1692 capsomeres arranged in a T = 169 skew icosahedral lattice. The genome is dsDNA of 330 kb; contain 701 potential overlapping ORFs (open reading frames). The proteins encoded include replication enzymes (DNA polymerase, DNA ligase and endonucleases), nucleotide metabolism enzymes (ATPase, thioredoxin,

Table 3.1 Floating Genera of plant viruses.

Genus	Type species	Particle shape and size	Genome type	Genome size	Capsid protein type and size	Others
Tobamo virus	Tobacco mosaic virus (TMV)	Rod-shaped; 300–319 nm L; 8 nm dia	(+)-ssRNA	6.3–6.6 kb	Single; 17–18 kDa	Mechanical transmission
Tobra virus	Tobacco rattle virus	Rod-shaped; 20–22 nm dia; 180–215 nm L	(+)-ssRNA 2 species	6.8 kb L, 1.8–4.5 kbS	Single; 22–24 kDa	Natural vectors are Trichodorid nematodes
Potex virus	Potato virus-X (PVX)	Flexuous, filamentous 470–580 nm L 13 nm dia.	(+)-ssRNA 5´ capped 3´ poly(A)	5.9–7.0 kb	18–27 kDa	Mechanical transmission
Carla virus	Carnation latent virus (CLV)	Flexuous, filamentous; 610–700 nm L; 12–15 nm dia.	(+)-ssRNA 3´ poly(A)	7.4–8.5 kb	31–36 kDa	Transmission by aphids, CPMMV by whitefly *Bemisia tabaci*
Allexi virus	Shallot virus X	Flexuous, filamentous ; ~800 nm L ; 12 nm dia.	(+)-ssRNA	9.0 kb	28–36 kDa	Transmission by eriophyid mites

(Contd.)

Table 3.1 *Contd.*

Genus	Type species	Particle shape and size	Genome type	Genome size	Capsid protein type and size	Others
Capillo virus	Apple stem grooving virus (ASGV)	Flexuous, filamentous ; 640–700 nm L ; 12 nm dia.	(+)-ssRNA 3´ poly(A)	6.5–7.4 kb	24–27 kDa	Seed transmitted
Fovea Virus	Apple stem pitting virus	Flexuous, filamentous ; ~800 nm L 12 nm dia.	(+)-ssRNA 3´ poly(A)	8.4–9.3 kb	28–44 kDa	No known vector
Tricho virus	Apple chlorotic leaf spot virus	Flexuous rod-shaped; 640–760 nm L; 12 nm dia.	(+)-ssRNA 3´ poly(A)	7.5 kb	22–27 kDa	No natural vector known
Viti virus	Grapevine virus_A	Flexuous, filamentous ; 725–825 nm L 12 nm dia.	(+)-ssRNA 5´ capped 3´ poly(A)	7.6 kb	18–22 kDa	GVA and GVB are transmitted by mealybugs (genera *Pseudococcus* and *Planococcus*), GVA also by scale insect *Neopulvinaria innumerabilis*

Furo virus	Soil-borne wheat mosaic virus (SBWMV)	Rigid rods 140–160 nm L, 260–300 nm L; 20 nm dia.	(+)-ssRNA Bipartite	6–7 kb L 3.5–3.6 kbS	19–21 kDa	Transmitted by fungus *Polymyxa graminis*.
Peclu virus	Peanut clump virus	Rod-shaped; 21 nm dia. 245 nm, 190 nmL	(+)-ssRNA Bipartite	5.9 kb L; 4.5 kbS	23 kDa	Transmitted by fungus *Polymyxa graminis* and by seed
Pomom virus	Potato-mop-top virus (PMTV)	Rod-shaped; 18–20 nm dia; 290–310 nm L, 150–160 nm L, 65–80 nm L	(+)-ssRNA Three molecules	6 kbL ; 3–3.5 kbL ; and 2.5–3 kbL	20 kDa	Infection restricted to dicots, transmitted by fungi; *Spongospora subterranea* & *Polymyxa betae* as vectors of PMTV & BSBV
Benyvirus	Beet necrotic yellow vein virus (BNYVV)	Rod-shaped; 20 nm dia; 390 nm L, 265 nm L, 100 and 85 nm L	(+)-ssRNA four molecules 5´ capped 3´ poly(A)	6.7 kb L, 4.6 kb L, 1.8 and 1.4 kb L	21–23 kDa	Transmitted by fungus *Polymyxa betae*

(Contd.)

Table 3.1 Contd.

Genus	Type species	Particle shape and size	Genome type	Genome size	Capsid protein type and size	Others
Hordei virus	Barley stripe mosaic virus	Rod–shaped; 20 nm dia; 110–150 nm L	(+)-ssRNA three molecules α, β and γ	3.7–3.9 kb α 3.1–3.6 kb β 2.6–3.2 kb γ	17–18 kDa	Transmitted by seed
Sobemo virus		Isometric (T = 3)	(+)-ssRNA Single molecule 5´ VPg	4.1–4.5 kb L	26–30 kDa	Transmitted by seed, beetle and myrids
Marafi virus	Maize rayadofino virus	Icosahedron; 28–32 nm dia; 2 components, B– contain genome and T– contain no RNA	ssRNA 5´ capped 3´ poly(A) with high cystidine	6.5 kb L	24–28 kDa minor and 21–22 kDa major protein	Restricted to Graminae, transmitted by leafhoppers
Tymo virus	Turnip yellow mosaic virus	Isometric (T = 3); ~30 nm dia; 2 components, B– contain genome and T– contain no RNA	(+)-ssRNA Single molecule	6.3 kb L	20 kDa	Restricted to dicots, transmitted by beetles of the familiesChryso-melidae and Curculionidae

Idaeo virus	Raspberry bushy dwarf virus (RBDV)	Isometric; ~33 nm dia.	(+)-ssRNA three molecules	5.5 kb, 2.2 kb and 1kb size	30 kDa		Restricted to *Rubus* species, transmitted by seed and pollen
Ourmia virus	Ourmia melon virus	Bacilliform; 18 nm dia and; 62 nmL, 46, 37, 30 nm L	Linear (+)-ssRNA 3 segments	2.9, 1.1 & 1 kb size	25 kDa		No natural vectors known
Umbra virus	Carrot mottle virus		Linear (+)-ssRNA	4 kb L	Helper virus of family *Luteoviridae* form coat protein		Transmitted by aphids in association with helper virus

ribonuclease reductase), transcription factors (TFIIB, TFIIS, RNase III), and enzymes involved in protein synthesis, modification and degradation (ubiquitin C-terminal hydrolase, translation elongation factor (EF-3), 26S protease subunit).

Small algal viruses   Virus infecting eukaryotic algae *Chara australis*, CAV, was described in 1976. CAV virus has rod shaped particles with a large genome (11 kb) than TMV (6.4 kb). The coat protein has similarities with that of BNYVV and TRV virus. The virus has homology of GDD-polymerase motif to BNYVV and nucleotide binding motif to that of potexviruses.

Cyanophages and Mycophages

Cyanophages are the viruses that attack cyanobacteria, i.e., members of the blue-green algae. Morris (1963) isolated a virus for the first time from the waste stabilization pond of Indiana University, USA, which attacked and destroyed the three genera viz., *Lyngbya*, *Plectonema* and *Phormidium*. Named after the genera infected these viruses are termed as LPP using the first letter of the genera. Several serological strains of *LPP* were isolated and named as *LPP-1*, *LPP-2*, *LPP-3*, *LPP-4* and *LPP-5*. They are commonly called as blue-green algal viruses or cyanophages.

Similarly phages infecting *Synecococcus elongates* and *Microcystis aeruginosa* are termed as SM group of phages (strains SM-1, SM-2) AS-1 for those infecting *Anacystis nidulans* and *Synecococcus cedrorum*, and N group of phages for those infecting *Nostoc muscorum*.

The LPP-1 and -2 , N-1 and SM-1 group of phages possess icosahedral head and a tail, while AS-1 has hexagonal morphology. They have a wide range of pH stability of 5–11 and temperature range of 4–40°C. They get inactivated mostly at 55°C. The growth cycle of these cyanophages includes a latent period of 6–8 minutes for LPP-1, -2, N-1, and AS-1 group of phages and about 32 minutes for SM-1 group of phages. Burst size ranges from 50 in case of AS-1 group to 300 for LPP group of phages.

Mycophages are viruses associated with fungi, also called as mycoviruses. Hollings (1962) has shown that viruses cause die back disease in mushrooms of *Agaricus bisporus*. The characteristic feature includes loss of crop and degeneration of mycelium in the compost. So far ~ 5000 fungal species are known to contain mycoviruses, most species of *Penicillium* and *Aspergillus* have been found to be infected with viruses. The mycoviruses have heterogenous properties with a diameter ranging from 25–50 nm and particle weight from 6–13×10^6 daltons. They possess 1–8 segments of dsRNA with a total molecular weight of 2–8.5×10^6.

DISEASE SYMPTOMS OF PLANT VIRUSES

Virus infection does not necessarily cause disease at all times in all parts of an infected plant. Symptoms may not be obvious in infected plants if:

- a very mild strain of the virus infects a plant;
- host is tolerant;
- a non-sterile 'recovery' from disease symptoms in newly formed leaves is made;
- age and position of leaves on the plant facilitates escape infection;
- the plant is infected with cryptic viruses;
- presence of dark green areas in a mosaic pattern.

Macroscopic symptoms

This comprises of both local and systemic symptoms which are stated as follows.

Local symptoms   Localized lesions that develop near the site of entry in leaves and are not of any economic importance but are important in biological studies. Infected cells may lose chlorophyll and other pigments giving rise to *chlorotic local lesions* (figure 3.1) that may be almost white or a very faint paler shade of green than the rest of the leaf.

(a)   (b)

Figure 3.1   Chlorotic local lesions in *Chenopodium quinoa*. (a) Coarse chlorotic lesions turning necrotic caused by BYMV. (b) Fine chlorotic lesions induced by GRV.

Many virus–host interactions where the infected cells die gives rise to *necrotic lesions* varying from small pinpoint areas to large irregular spreading necrotic patches (figure 3.2).

Figure 3.2   Necrotic local lesions in *Nicotiana tabacum* induced by TMV.

In a third type, *ringspot lesions* appear, which consists of a central group of dead cells. Beyond this, there develop one or more superficial concentric rings of dead cells with normal green tissue between them (figure 3.3). Some viruses in certain hosts show no visible local lesions in the intact leaf, but when the leaf is cleared in ethanol and stained with iodine, 'starch lesions' may become apparent.

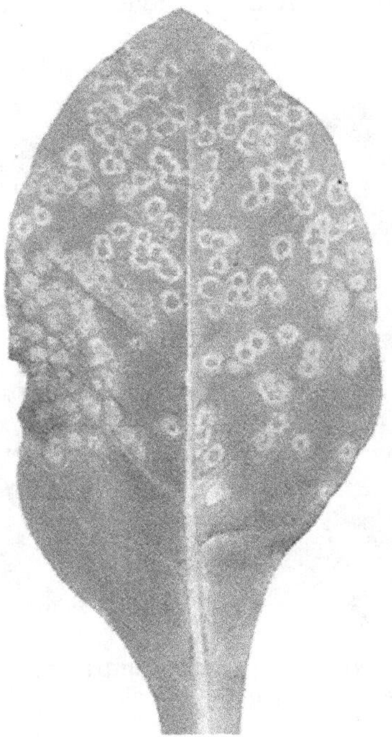

Figure 3.3   Necrotic ringspots in inoculated leaf of *Nicotiana tabacum* caused by ToRSV.

Systemic symptoms

1. *Effects on plant size*   Reduction in plant size (stunted growth) induced by virus infection where the systemically infected plant shows no obvious sign of disease. For example, mild strains of PVX infecting potatoes in the field may cause no symptoms, which may lead to reduced tuber yield by about 7–15% known through carefully designed experiments. The degree of stunting is in proportion to the severity of symptoms.

2. In perennial deciduous plants such as grapes, there may be a delayed initiation of growth in spring. Root initiation in cuttings from virus-infected plants may be reduced, as in chrysanthemum. Stunting may affect all parts of the plant. A reduction in total yield of fruits is a common feature and an important economic aspect of virus diseases.

Figure 3.4 Systemic mosaic symptoms of TMV infection in *Nicotiana tabacum.*

3. *Mosaic patterns* Development of a pattern of light and dark areas gives a mosaic effect in infected leaves. Nature of the pattern varies with different virus–host combinations. In dicot plants, the mosaic have generally irregular outline with dark and pale or yellow-green colour as seen in TMV-infected tobacco leaves (figure 3.4) or with many different shades as seen with TYMV in Chinese cabbage (figure 3.5). The virus moves up from the inoculated leaf into the growing shoot and partly into expanded leaves leading to "clearing" or "yellowing" of the veins. Vein banding may persist as a major feature of the disease.

4. The development of the *stripe* diseases found in monocot plants follow a similar pattern to that found for mosaic diseases in dicots.

5. *Yellow disease*   Viruses that cause a general yellowing of the leaves as seen with yellows in sugar beet, are of

Figure 3.5   Chlorotic mosaic induced by TYMV in Chinese cabbage.

considerable economic importance. The first sign of infection is usually a clearing or yellowing of the veins in the younger leaves followed by a general yellowing of the leaves. No mosaic is produced. When severe, a yellow disease may lead to a total loss of the crop.

6. *Leafrolling*   Virus infection can lead to leafrolling, which is usually upwards figure 3.6 or occasionally downwards.

7. *Ringspot diseases*   Many viral diseases produce a pattern of concentric rings and irregular lines on the leaves figure 3.7 and sometimes also on the fruit. The lines may consist of yellowed tissue or may be due to death

Figure 3.6   Leafrolling induced by BLRV in broad bean.

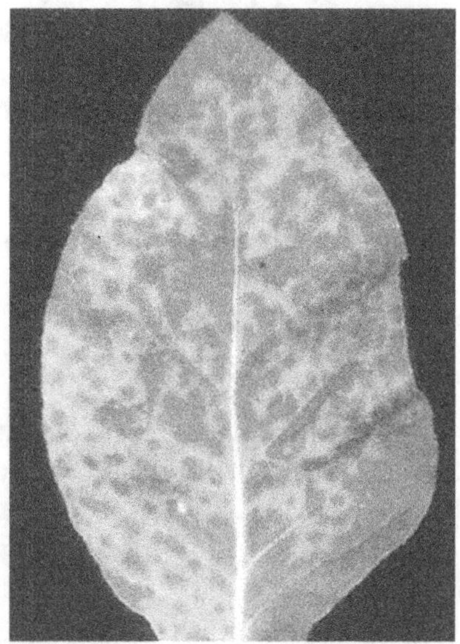

Figure 3.7   Systemic ringspotting caused by TMV in *Nicotiana clevelandii*. Note that the ringspots are spreading and fusing with one another.

of superficial layers of cells, giving an etched appearance. In severe diseases, complete necrosis of the leaf lamina may occur, such as TRSV. Ringspot patterns may also occur on other organs, for example, bulbs and the necrotic ringspot of tubers by PVY, often around the eyes, that become sunken and necrotic.

8) *Necrotic disease* Death of tissues, organs or the whole plant is the main feature of some diseases. Necrotic patterns may follow the veins as the virus moves into the leaf (figure 3.8). In some diseases, the whole leaf is killed. Necrosis may extend rapidly throughout the plant as observed with PVX and PVY infection of potatoes, where necrotic streaks appear in the stem. Necrosis spreads rapidly to the growing point which is killed and subsequently all leaves may collapse and die.

Figure 3.8   Tomato streak disease (ToMV + PVX) causing severe necrosis in a tomato plant that would subsequently die.

9.   *Wilting*   Wilting of the aerial parts frequently followed by death of the whole plant may be an important feature as seen with virus diseases of chickpea.

10. *Developmental abnormalities*   Uneven growth of leaf lamina in mosaic disease, swellings in the stem of woody plants such as those seen in cocoa swollen shoot disease, outgrowths from the leaf surface usually associated with veins called enations, tumour–like growths. Some consist of wart–like outgrowths on stems or fruits (seen with WTV virus), and stem deformation (stem splitting and scar formation) in some woody plants. One of the unusual symptoms of BSV in some Musa cultivars is that the fruit bunch emerges from the side of the pseudostem instead of from the top of it, which is due to necrosis of the ciga leaf.

11. *Reduced nodulation*   Reduction in the number, size and fresh weight of nitrogen–fixing Rhizobium nodules induced by virus infection in legumes as seen with AMV in alfalfa.

12. *Genetic effects*   Infection with BSMV induces an increase in mutation rate in Zea maize and also a genetic abnormality known as aberrant ratio (AR). More than one phenomenon is involved in the genetics of the AR effect.

Agents inducing virus-like symptoms include nutritional deficiencies, high temperature, hormone damage, insecticides, bacteria, toxins produced by arthropods, and small cellular parasites such a mycoplasmas (now termed as phytoplasmas), spiroplasmas and rickettsia-like organisms.

All plant viruses have the following common properties. Some plant viruses have a very limited host range and others attack numerous species.

- Plant viruses can multiply only within living cells.
- Plant viruses usually multiply only within living plant

cells, but some may be able to multiply within the bodies of aphids and nematodes.

- A given plant virus may be able to multiply only within the living cells of one species or genera of plants, but some can multiply within the cells of a wide group of plant families.

- Whether or not latent plant viruses exist and how to identify such viruses is still dogmatic. In general, most organisms harbour both virulent and latent viruses.

- Plant viruses do not attack animals and vice versa.

- Temperate viruses embed themselves within the hosts nucleic acid and are transmitted from one generation to the next just like genes of the host.

- Often viruses reside in their host, without causing any disease or symptom. Such latent viruses are undiscovered.

TRANSMISSION OF PLANT VIRUSES

Plant viruses vary in their mode of transmission. Usually only a single mode of transmission is important, but some viruses are transmitted by more than one mode.

- Aphids or other sap-suckers are the most common transmitters.

- Nematodes living in the soil transmit some viruses.

- Sap transmission is important only for a few viruses and occurs on cultivators, pruning, hands of workers and clothing of workers. These are important for some potato viruses.

- Pollen transmission from a male flower to the female occurs for a few viruses. Such viruses are seed-borne.

- Vegetative propagation and grafting.

- Seed, pollen, mites, nematodes, dodder, fungi (carried by zoospores and mycelium) and insects (aphids, leafhoppers, scale insects, thrips, grasshoppers, beetles, whiteflies). For example, cucumber mosaic virus and barley yellow dwarf virus are transmitted by aphids.

Stages in Transmission Cycle

There are three stages in the transmission cycle.

1. Acquisition stage—vector feeds on infected plant and acquires sufficient virus for transmission.
2. Latent period—vector has acquired virus but is not able to transmit it. For externally borne virus there is little or no latent period.
3. Retention (transmission) period—Length of time during which vector can transmit the virus to a healthy host.

Non-Persistent Transmission

Virus genera involved in this type of transmission include Alfamovirus, Caulimovirus, Cucumovirus, Potyvirus, Macluravirus and Fabavirus. Acquisition time is rapid in matter of seconds, retention time with a non-persistent virus aphid begin to lose its ability to infect immediately after acquisition. The rate at which infection is lost depends on many factors including temperature. *Myzus persicae* is known to be able to transmit a large number of non-persistent viruses.

Semi-Persistent Manner

Properties that are intermediate between non-persistent and circulative systems, and requires helper components in the form of virus encoded proteins for example, Caulimoviruses. MCDV is transmitted in this manner by vector *Graminella nigrifrons* and RTSV by *Nephotettix virescens*. Acquisition time and inoculation time includes 15 min. each for MCDV, and 5–30 and 5–10 for RTSV. AYV and PYFV are transmitted by aphid *Cavariella aegopodii* in which AYV itself act as a helper part for PYFV. Similarly, P2 and P3 proteins of 18 kDa and 15 kDa respectively of CaMV are helper proteins in aphid transmission.

Bimodal

Virus transmitted by both non-persistent and semi-persistent manner by the same aphid are termed as bimodal.

Persistent

Usually involve one or a few species of aphids. Viruses that are internally borne in their aphid vectors may replicate in the vector (propagative) or may not (circulative); in both cases virus enters the salivary gland after ingestion by the aphid. Yellowing and leaf rolling are commonly produced by infection with persistent transmission of virus.

Circulative (persistent) viruses   include Luteoviruses among which BYDV and PLRV are the most studied. For BYDV, acquisition time is as little as 5 min to several hours, the latent period is 12 hrs, after which the virus can be transmitted with an inoculation time of about 10–30 min. by vector *Schizaphis graminum*. Umbravirus vectors include *Cavariella aegopodii* for carrot mottle virus, *Aphis craccivora* for Groundnut rosette virus, *Myzus persicae* for tobacco mottle virus, *Acyrthosiphon pisum* for Bean yellow veinbanding and Pea enation mosaic virus.

Vectors

Many plant viruses are transmitted from plant to plant by invertebrate vectors mainly by members of phylum Arthropoda (Urinamia and Crustacea) and Nematoda.

Insecta which is a Subphylum of Urinamia contains a number of vector species (Orthoptera, Dermaptera, Coleoptera, Lepidoptera and Diptera), most of which are chewing insects that feed on green plants. Thysanoptera (Thrips) are rasping and sucking plant feeders and those of Hemiptera feed by sucking on green plants.

Arachnida   Order *Acari* which includes mites and ticks contain members that feed on green land plants. *Acari* has two families Tetranychidae and Eriophyidae that are known viral vectors.

Nematoda   Order Tylenchida are parasites living on green plants but are not viral vectors. Vectors known so far are confined to the Dorylaimida group.

Aphids (Insecta)   *Myzus persicae* is an important vector aphid. Its mouthparts consist of two pairs of flexible stylets held within the groove of labium. The maxillary stylet has a series of tooth-like projections near their tip. During feeding a drop of gelling saliva is secreted. The stylets then rapidly penetrate the epidermis. Maxillary stylets usually move between cells until they reach a phloem tube. Aphid transmission is mainly by cell injury. Unstable viruses are readily transmitted by aphid feeding, but some stable viruses like TMV, TYMV and SBMV do not get readily transmitted.

Leafhoppers and Planthoppers (Auchenorrhyncha)

Eggs of these hoppers hatch to nymphs, which feed by sucking leaves and pass through a number of moults before becoming an adult. Leafhoppers are similar to aphids in their mouthparts surrounded by salivary gland and penetrating to the phloem in preferred hosts. *Graminella nigrifrons* and *Nephotettix virescens* hoppers transmit viruses MCDV and RTSV respectively in a semi-persistent manner as discussed before. Two genera of Gemini viruses (mastrevirus and curtovirus) are transmitted by leafhoppers in a persistent (circulative) manner. BCTV, MSV and WDV are transmitted by *Circulifer tenellus*, *Cicadulina mobila* and *Psammotettix alienus* respectively. RDV and RRSV are transmitted in a persistent propagative manner by *Nephotettix cincticeps* and *Niloparvata lugens* respectively.

Whiteflies

Three genera of viruses are transmitted by whiteflies viz. the Begomoviruses, Criniviruses and some Closteroviruses. Begomovirus by *Bemisia tabaci* vector, Clostero and criniviruses (BPYV, TICV and CCSV) by glasshouse whitefly *Trialeuroides vaporariorum*.

Thrips

Ten species of Thripidae family are vectors of plant viruses. Thrip feeding apparatus consist of one mandible that punches a hole in the leaf and two maxillae that are inserted into the plant cell through which cell contents are sucked. *Frankiniella occidentalis*

Thrip species	Virus family	Virus
Frankiniella occidentalis	Tospovirus	GRSV, INSV,
	Ilarvirus	PDV
F.fusca	Tospovirus	TSWV
Thrips australis	Ilarvirus	PNRSV
T.palmi	Tospovirus	GBNV, WSMV
T.tabaci	Sobemovirus	SoMV
	Ilarvirus	PNRSV, TSV

was the most efficient vector for four tospoviruses TSWV, INSV, TCSV and GRSV. Median acquisition time found to be 106 min. and the median inoculation time to be 58 min. Details of vectors and virus transmitted are stated in the table above.

Mealy Bugs

They are much less mobile than aphids, phloem feeders, vectors of Badnavirus and Terovirus (LCV, PMWaV). CSSV transmitted by mealy bug *Pseudococcus njalensis* takes 16 min. to penetrate leaf and virus persists for 3 hrs in the bug. PMWaV transmitted by pink pineapple mealybug *Dysmicoccus brevipes*. GLRaV–3 transmitted semi-persistently by *Planococcus citri* and the virus retained in vector for 34 hrs.

Mites

Members and families of Eriophyidae and Tetranychidae feed by piercing plant cells and sucking the contents, act as vectors for Rymoviruses. *Aceria tulipae* transmits WSMV acquisition time and inoculation time being 15 min. each and retain the infectivity for 6-9 days. *Eriophyes tulipae* transmits WSMV in a circulative manner. *Petrobia latens,* a brown wheat mite transmits BaYSMV.

Pollinating Insects

RBDV and other viruses are transmitted through infected pollen and BLMV via pollen by foraging honey bees.

Nematodes

Viruses transmitted through soil by nematodes inclusive of two genera of plant viruses. Nepoviruses are transmitted by *Xiphinema* spp. and *Longidorus* spp., and Tobraviruses (PEBV, TRV) by *Trichodorus* and *Paratrichodorus* spp.

Fungi

Vectors are members of class Plasmodiophoromycetes (*Polymyxa* and *Spongospora*) that transmit rod shaped or filamentous viruses, and class Chytridiomycetes (*Olpidium brassicae* and *O. bornavirus*) transmit viruses with isometric particles. In vitro fungal transmission occurs between *Olpidium* species and isometric viruses of Tombusviridae. Whereas, in vivo transmission occurs between rod shaped viruses of the Bymovirus, Furovirus and Varicosavirus genera and *O. brassicae* and the three Plasmodiophoral species: *O. brassicae* and LBVV; *P. graminis* and SBWMV; *P. betae* and BNYV V.

DETECTION OF PLANT VIRUSES

Until a virus is detected, its presence is not known. Nearly everyone who received an early polio vaccine is now a carrier of Simian virus 40 (SV40). The monkey cells used to grow that vaccine were infected by SV40 but no one knew that SV40 existed. Many plants are carriers of plant viruses but these viruses hardly cause any disease. The yield might be reduced but that would not be visible either.

The methods for detecting plant viruses include:

1. *Use of antibodies to the virus*   In the ELISA assay, plant sap is placed in a plastic tray which contains wells. Proteins including the virus adsorb to the well. Then the tray is washed under running water. An antibody to the virus is added. A special enzyme has been bonded to the antibody. Very little of this antibody-enzyme binds to the well because all the plastic of the tray is covered with protein. Wash again. Now a colourless dye complex is formed which the enzyme can split. The dye that is split free has colour. The amount of colour correlates with the amount of virus present in the donor plant.

2. Graft a leaf from the suspected plant to an indicator plant. The virus moves into the host plant and causes symptoms in the indicator plant.

3. *Sap transmission*   Place a drop of sap from the suspected plant on an intact leaf of an indicator plant. Add some grit to the sap and rub so that the leaf is scratched and the virus, if any, can get into the leaf. If the indicator plant shows symptoms then we know that the virus came from the donor plant. The indicator plant for sweet potato viruses is Brazilian Morning glory, tobacco, tomato, lambsquarter, and other plants that are used as indicators in the sap and grafting assays.

4. Visual inspection with EM.

5. By eliminating the possibility that symptoms are not due to other sources (e.g. herbicide, nutritional deficiencies).

Creating Virus-Free Clones of Plants

It is not possible to cure a plant affected by viruses, but one of the following methods may give a virus-free clone. Since the virus-free plant might contain a virus you did not know about, it is proper to call them virus-tested or virus-indexed plants.

- Often the terminal bud of a plant is free of virus. Remove the 0.03 to 0.05 mm bud tip and grow it in sterile agar containing sucrose and everything that the bud needs. 70 to 99.9% of the buds will rot or fail. If some grow into plants, test those plant borne viruses as above and you might find one which is free of virus.

- Some viruses infect the tip also and the above method will not work. In such cases try growing the plant in an incubator that is so hot that the plant barely grows. Often you must supply extra carbon dioxide at a very precise level. Then try to start a new plant as above and test it for all the known viruses.

- Most viruses are not seed-borne. Therefore, a new seedling may be free of virus. Test the seedling for all known viruses.

- Treat your pruning tool with viricide between plants. Use one volume of Chlorox plus nine volumes of clean water. Milk inactivates many viruses—use milk to wash tools/hands.

- Removing diseased plants, killing and removing potential virus vectors (primarily weeds and insects).

- Disease-resistant cultivars, disease or virus-free seed, roots or tubers.

- Cross protection (inoculation with a less-virulent strain of a virus protects the plant from a more virulent strain later when exposed to it).

- Heat (some viruses are killed at temperatures that will not kill the host). For example, dormant propagative organs dipped in hot water (35°C) for few minutes or hours, or growing plants in greenhouse at 35–40°C for several days, weeks or months may inactivate virus.

Once you have a virus-indexed, plant keep it in a screened cage where it cannot be infected via a vector. Alternatively, keep it on sugar agar in a sterile glass jar under fluorescent light at a low temperature where it grows slowly. Use scions from this microplant to grow plants for the field.

> ## Websites about Plant Viruses
>
> *Http:// image.fs.uidaho.edu/vide/descr267.htm*
>   lists plant viruses.
> *Http:// image.fs.uidaho.edu/vide/famly050.htm*
>   lists the species of plants which are susceptible
>   to the virus you choose.

PLANT DISEASES CAUSED BY VIRUSES—A HISTORICAL REVIEW

a. Plant viruses consist of a nucleoprotein that multiplies only in the living cells of a host. The presence of viruses in host cells often results in disease.

b. 400 or more viruses are known to attack plants (2000 viruses are described for plants, animals, bacteria, etc.). Viruses are generally specific, what infects a plant does not cause disease in an animal, and *vice versa*.

c. The first record of a disease that was later found to be caused by a plant virus was on tulips in the 17th century in the Netherlands.

d. The first experimental demonstration of the infectious nature of viral disease was recorded by Lawrence, who described the transmission of a disease of jasmine by grafting.

e. Adolf Mayer (1886) described a disease of tobacco called *mosaikkranheit* (tobacco mosaic). This disease could be transmitted to healthy plants through the sap from diseased plants.

f. Dmitrii Ivanovski (1892) demonstrated that the agent in tobacco mosaic was filterable. He demonstrated that the causal agent of tobacco mosaic could pass through a filter that retains bacteria.

g.   In 1898, Martinus Beijerinck demonstrated that the causal agent was not a microorganism but a *contagium vivum fluidum* (contagious living fluid). He was the first to use the term *virus*, which is the Latin word for poison. He concluded that this was not a toxin, because repeated inoculations of diluted infected sap caused similar disease as it was passed from one plant to another. If it had been a toxin, it would eventually be diluted away.

h.   Loefler and Frosch (1898) described the first filterable infectious agent in animals, the foot-and-mouth disease virus and Walter Reed (1900) described the first human virus, yellow fever virus.

i.   In 1929, F. O. Holmes provided a tool by which the virus could be measured by showing that the amount of virus present in a plant sample preparation is proportional to the number of local lesions produced on appropriate host plant leaves rubbed with the contaminated sap.

j.   In 1935, W. M. Stanley isolated and purified some tiny white crystals from leaves of mosaic-infected tobacco plants. He treated healthy plants with TMV, which had been precipitated out of infected tobacco juice with the help of ammonium sulphate and a technique he had developed. The healthy plants contracted tobacco mosaic disease. Due to the high protein content of the purified virus particles, he concluded that the virus was an autocatalytic protein that could multiply within living cells. Although his conclusions were later proved incorrect, Stanley received the Nobel Prize in 1946 for this work..

k.   In 1937, Bawden and Pirie demonstrated that virus consists of protein and nucleic acid (RNA).

l.   In 1939 Kausche saw virus particles for the first time with the electron microscope.

m.   During 1955–1960 much was learned by various workers, regarding the infectivity of viral (TMV) RNA and the structure and arrangement of viral (TMV) coat protein.

n.   In 1971, T. O. Diener discovered viroids, which only consist of nucleic acids. Smaller than viruses, these caused potato spindle tuber disease (250–400 bases long of single-stranded circular molecule of infectious RNA). About a dozen other viroids that cause disease in a variety of plants have been isolated. No viroids have ever been found in animals.

o. In 1900, cauliflower mosaic virus, whose genome is a circular double-stranded DNA chromosome, was the first plant virus for which the exact sequence of all its 8000 base pairs was determined. In 1982, the complete sequence of the bases in the single-stranded tobacco mosaic virus RNA was determined, as were those of smaller viral RNA and of viroids.

p. In 1986, use of transgenic plants to obtain resistance against viruses (TMV) was obtained.

VIRUS DISEASES OF PLANTS—
SYMPTOMS SPECIFIC

a. The symptoms of specific plant diseases form the basis for the following disease names: tobacco mosaic, turnip crinkle, barley yellow dwarf, ringspot of watermelon, cucumber mosaic, spotted wilt of tomato.

b. Some viruses have a broader host range than the name of disease or virus may imply. For example, tobacco mosaic virus (TMV) infects tomato, eggplant and pepper, in addition to tobacco.

PROPERTIES AND MORPHOLOGY
OF PLANT VIRUSES

1. Plant viruses are non–cellular, ultramicroscopic particles that multiply only in living cells. Their size is measured in nanometres.

2. Most plant viruses consist of protein shells surrounded by a core of positive-stranded nucleic acid (normally ssRNA—nucleotides (guanine, uracil, cytosine, adenine) + 5 carbon sugar called ribose + a phosphate group—but sometimes these viruses contain dsRNA or dsDNA

(2 strands of nucleotides with thymine substituted for uracil and deoxyribose instead of ribose).

3. 5–40% of virus is made up of nucleic acid and 60–95% is composed of protein.

4. Protein coats or shells can be of different shapes, but are normally rod, filamentous, isometric, quasi-isometric/ bacilliform or variants of these structures. For example, tobacco mosaic and barley stripe mosaic viruses are rods, while broad bean wilt and maize chlorotic dwarf viruses are isometric or more spherical in shape.

VIRUS GENOME

Minimum number of genes in a plant RNA virus could be two: a coat protein and an RNA replicase gene (as is the case with RNA phages). Evidence indicates there are usually 3–5 gene products.

Plant *positive-stranded RNA* viruses frequently possess divided genomes. In addition, viral genomes are separately encapsulated. Viral genomes consisting of two or three different nucleic acid components, all required for infection are called bipartite, tripartite or multipartite viruses. Multipartite viruses are potentially at an evolutionary disadvantage. Infectivity dilution curve for Alfalfa mosaic virus (requiring B, M, Tb particles for infectivity) is steeper than for tobacco necrosis virus (single particle). Partition of genome could potentially hinder transmission or infection by a virus.

SATELLITE VIRUSES AND RNAs

Kasinis in 1962, described the first satellite viruses. These viruses are serologically unrelated to their helpers and the two genomes exhibit little if any sequence similarity. Satellite viruses are dependent for its replication on the presence of a second, independently replicating virus.

Satellite RNAs have no coat protein of their own and are encapsulated with the help of other viral RNAs.

PROTOTYPE VIRUSES

I. Tobacco Mosaic Virus

Tobacco Mosaic disease is caused by Tobacco Mosaic Virus (TMV). This virus has worldwide distribution and primarily infects tobacco and tomato, but more than 350 species are susceptible. The different characteristics of this plant virus are as follows:

1. TMV is a rod-shaped particle which is 300 nm long 15–18 nm in diameter (figure 3.9). It possesses ssRNA and a protein coat.

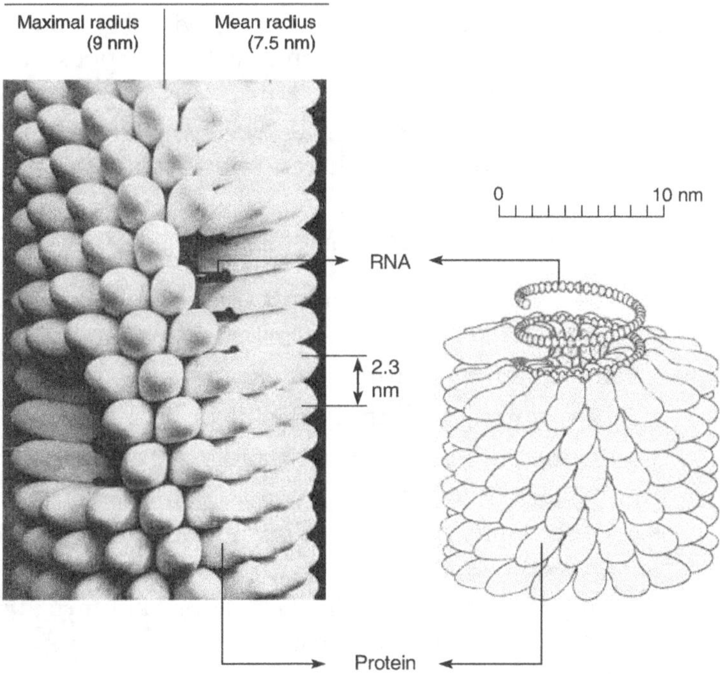

Figure 3.9   Electron microscopy picture of tobacco mosaic virus (TMV).

2. TMV is difficult to inactivate, and can survive for almost 5 years in dead, dried tissues and for many months in living plant tissues.

3. There are many strains of the virus that vary in their virulence causing severe to mild symptoms. Virus is spread from plant to plant through injuries caused by crop worker, contaminated equipment and chewing insects.

4. These viruses are found in dead plant tissues and debris, on contaminated equipment, in contaminated soil, greenhouse containers, bedding, tools, and in living hosts, including weeds like horsenettle, *Solanum carolinense*, and other crop plants (tomato, pepper, and eggplant).

Symptoms of the TMV disease are:

a. Tobacco leaves become mottled with light and dark green areas; leaves become distorted; puckering or blistering takes place especially in areas of new growth.

b. In tomato, mottling of leaves occurs and leaflets become long and pointed and there is stunting of plant growth.

Management of Tobacco Mosaic Disease

*   Use virus-free seed (tomato seed can be treated with acid or bleach).

*   Transplant in non-infested soil.

*   Fumigate with methyl bromide or heat.

*   Prevent chewing of tobacco or smoking around seedbeds or in greenhouses.

*   Wash hand with soap and water or milk to eliminate spreading of virus.

*   Spray plants with milk (whole or skimmed) which seems to help reduce infection.

*   Practise crop rotation with nonhost crops (corn, rice and other cereal grains).

*   Use resistant cultivars.

Structure and Self-assembly

The stability of the TMV particle accounts for its having been the first virus to be identified, purified to homogeneity, and then biochemically and biologically characterized. The characteristic features of the physical structure includes:

1. This rod shaped virus is of length 300 nm and radius 9 nm, with a central hole of radius ~2 nm.

2. The subunits are arranged in a single helix which is right handed with $16\frac{1}{3}$ protein subunits per turn.

3. The ssRNA binds at a radius of ~4 nm with 3 nt per protein subunit and the length were either 6395 or 6398 nt (due to polymorphism near 5′ terminus), this corresponds to ~2140 protein subunits and 130 turns of the viral helix.

4. The main form of protein aggregates at high pH and low ionic strength is a mixture of monomer and small aggregates, collectively known as "A-protein". The two-layered disk of TMV protein is central among the aggregates.

5. Protein and RNA are more infectious than naked RNA alone (nearly 1000 times the amount of naked RNA is required to cause infection) and the viral RNA is the sole determinant of tobacco mosaic disease.

Self Assembly Self assembly involves two major steps, viz; nucleation and elongation. Nucleation involves stacking of subunits along the RNA strand and is highly selective for the TMV RNA species (very high specificity). Elongation or growth is bidirectional as observed in the Vulgare strain of TMV, but is faster in the 5′-direction and is ~5-fold. Aggregates of 33 protein molecules form the double disk. This combines with viral RNA. Attachment of the nucleic acid to the protein aggregate begins at a unique region on the viral RNA called as origin of assembly site (OAS) about 900 nucleotides from the 3′ terminus of TMV RNA. Co-translation disassembly is defined as the process in which the protein coat is displaced at the 5´ end by ribosomes in the host cell.

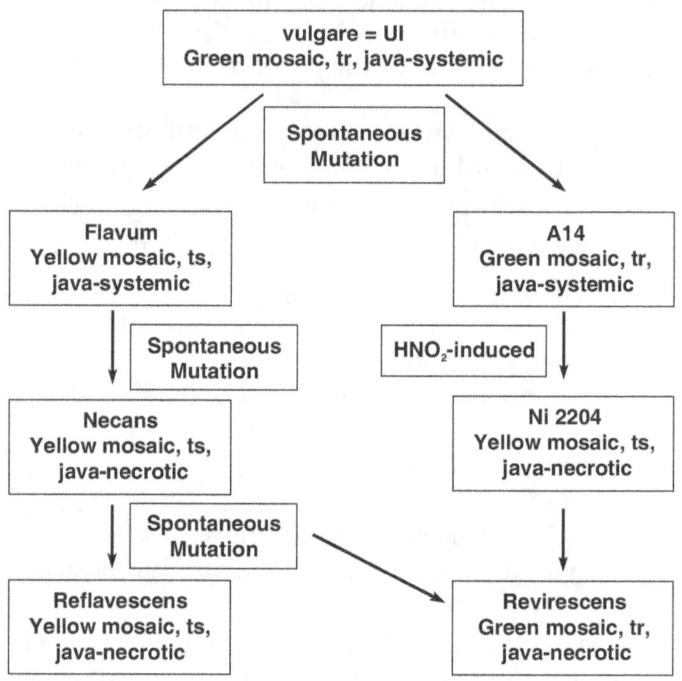

TMV Mutants

The various TMV mutants are as follows:

- Wild strain Vulgare (= U1) is the most commonly seen in grown tobacco.
- J14D1 is a variant of J14 which produces lesions on the Turkish tobacco but does not spread systematically, whereas J14D1 spreads systematically.
- Yellow aucuba strain (YA) produces necrotic lesions and spreads systematically, and is also temperature-resistant.
- Green aucuba strain (GA) produces similar symptoms to that of YA except that it is temperature-sensitive.

Antigenic Structure of TMV

Types of epitopes   There are two broad classes of epitopes:
1. Continuous (regions 1–11; 34–39; 62–68; 76–88-helix;
80–90; 103–112; 115–134-helix; 134–146; 149–158) and
2. Discontinuous ('A' epitope – 66,67, 140-143; 'B' epitope
– 1, 3, 4, 9, 150, 152, 153). *Neotopes* are epitopes specific
for the quaternary structure of virions of TMV, deduced
by using monoclonal antibodies. Cryptotopes are epitopes
absent in TMV particle, but exist in TMV protein subunits of
region 103–112 located in a central loop of polypeptide chain.
Virus detection is done by serology using double antibody
sandwich ELISA and immunodiffusion.

Infectivity

Viral ingress into the plant is by a small, healable wound. There
are no known vectors observed. The virus movement from
cell to cell occurs through plasmodesmata mediated by virus
coded 30 kDa protein. The long distance virus movement in
the host at a fast rate is through stem phloem tissue without
any replication. The symptom determinants include systemic
mosaic and hypersensitive necrotic local lesions, which are
determined by N-gene of the tobacco plant.

Replication

Gene organization   TMV RNA itself is an mRNA of 6395
nt, referred to as (+)-sense RNA, containing 5 genes. The
5´ proximal gene encodes two co-initiation proteins of 126
kDa and 183 kDa. The 3 genes located internally in the genome
are not expressed from genomic RNA but rather from
subgenomic (sgRNA) RNA generated during the course of
infection (figure 3.10). SgRNAs include I_1 mRNA of 54 kDa
(begins at 3405, 2991 residues), I_2 mRNA of 30 kDa (begins at
4838, 2058 residues) and LMC-RNA of 17.5 kDa (begins at
5703, 693 residues). Both genomic RNA and LMC-RNA have
been shown to be capped at 5´ end by m 7G. The 3´-end of

TMV RNA ends with the sequence –C-C-C-A and can be charged with an amino acid (histidine). This region is non-coding and can be folded into a tRNA-like structure preceded by a series of four pseudoknots.

Viral RNA contains:

- a 5´-UTR (untranslated region) of 68 nt (nucleotides);
- a 126 K gene readthrough of UAG termination site (3417-19) by plant suppressor Tyr-tRNA to provide 183K protein at UAA (4917–s19);
- a 54 K gene (3495) coincide with C-terminal 1/3 of 183 kDa protein;
- a 30 K gene begins at 4903 and overlaps 54K and 183K by five amino acids;
- a 5 nt internal non-translated region (5707–5711) which includes UAA terminal site of 30 K gene;
- a 17.5 K gene beginning at 5712 terminate with UGA at 6189-91; and
- a 3´-end UTR of 204 nt is also observed.

Proteins synthesized in vivo

- 183 kDa and 126 kDa proteins found in infected tobacco leaves are replicase proteins involved in viral RNA replication as RDRP (homology with reverse transcriptase of HIV).
- 30 kDa protein I_2 gene product acts as movement protein (MP) of TMV which ads in cell-to-cell movement.

TMV Life Cycle

Virus enters by force, and uncoating and exposure of 5´ORF of TMV RNA is facilitated by Ca^{2+} binding sites (~10–7 M intracellular Ca^{2+}). Uncoating is a bi-directional process using co-translational mechanism for the 5´ → 3´ direction. Disassembly and replication are coupled events. Three virus encoded enzyme activity seen in replication of TMV RNA viz., RNA-dependent RNA polymerase (RDRP) that synthesizes

(–)-RNA using an RNA template, a helicase involved in RNA displacing activity, and methyl transferase leads to 5′ capping of RNA. In protoplast (-)-strand synthesis ceases 6–8 h post inoculation, but (+)-strand synthesis continues for a further 10 h which are encapsidated into virions. 5′ORF of TMV encode 126 kDa protein, the stop codon of which is read through to

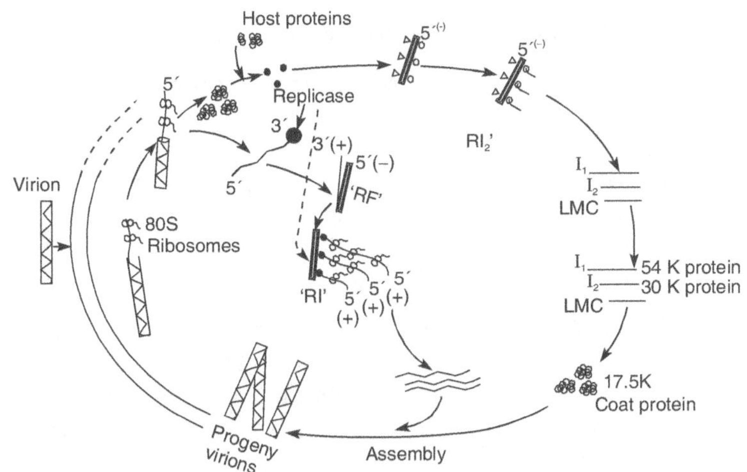

Figure 3.10 Tobacco mosaic virus (TMV) life cycle.

give a 183 kDa protein. Both of these proteins have helicase motifs and 183 kDa in addition has RDRP motif. Viral RNA binds a replicon (formed by combination of 126 K and 183 K proteins—active replicase) molecule at its 3′-end and become a template for (-)-strand synthesis. This structure is called "Replication form (RF)". 3′-UTR of TMV RNA can be folded into three structure domains; a 3′-domain mimicking a tRNA acceptor branch, an analogue of a tRNA anticodon and an upstream domain comprising 3 pseudoknots each containing 2 double helical segments. 30 kDa protein of TMV was identified as MP, which has the abilities to bind nucleic acid, to increase plasmodesmatal SELs, to facilitate movement of RNA to neighbouring cells as microinjection, and interacts with cytoskeletal elements (microtubules in the protoplasts and pectin methyltransferase (PME) of cell walls) of the cell. Rate of movement from upper to lower epidermis was found to be 8 μm/h.

Etiology Mild mosaic, yellow mosaic, white mosaic, leafburning type, stunting, leaf distortion and necrosis. Symptoms include mosaic of light and dark green areas on leaves, less common symptoms include leaf distortion, vein clearing in young leaves and necrotic areas in old leaves.

Cytopathological inclusions of TMV

- Crystalline inclusions in the cytoplasm stain green in the Calcomine Orange-Luxol brilliant green stain, and magenta in Azure-A after heating for 1 min. at 60°C.
- Paracrystalline inclusions as elongated aggregates of TMV particles located in vacuoles as a result of transformation of crystalline inclusions.
- X-bodies—coarse to fine granular bodies, rounded or amoeboid in outline, either vacuolated or non-vacuolated observed in cytoplasm and are termed as X-bodies. It is bounded by plasma membrane. It contains broad filaments, non-structural TMV protein which aggregates within electron lucent areas and host components.

Natural Host Range and Symptoms

Symptoms persist.

- *Nicotiana tabacum*—leaf mosaic, severe crop losses.
- Also found in many other plant species.

Transmission

Transmission does not involve a vector, but is by mechanical inoculation. Some of the methods are:

- transmission by grafting,
- transmission by contact between plants,
- transmission by seed (occasionally transmitted through the testa, but not through the embryo),
- not transmitted by pollen.

The diagnostically susceptible host species and symptoms are systemic mosaic in *Nicotiana tabacum* (various cultivars), local lesions in *Chenopodium quinoa, Nicotiana glutinosa* and *N. tabacum* cv. Xanthi-*nc*.

Families Containing Susceptible Hosts

a.   Chenopodiaceae (4/5), Compositae (1/6), Cucurbitaceae (3/3), Leguminosae–Papilionoideae (1/2), Papaveraceae (1/1), Solanaceae (15/19)

Families Containing Insusceptible Hosts

a.   Caryophyllaceae (2/2), Chenopodiaceae (1/5), Compositae (5/6)
b.   Convolvulaceae (2/2), Cruciferae (5/5), Gesneriaceae (1/1), Graminae (5/5)
c.   Labiatae (1/1), Leguminosae-Papilionoideae (1/2), Malvaceae (1 /1)
d.   Solanaceae (4/19), Umbelliferae (4/4)

Control

Prevention of primary infection is possible through the following controls.

- Elimination of soil-borne inoculum,
- Elimination of tobacco products,
- Control of inoculum from weeds and solanaceous crops,
- Destruction of inoculum on equipment.

Prevention of secondary infection is by:

- Rogueing—removal of infected plants from the field before each transplantation,
- Using chemicals such as Dodine and Glyodin,
- Using milk to coat plants that are infected.

CAULIFLOWER MOSAIC VIRUS (CaMV)

Nomenclature

The synonyms that are used in the naming of the virus are: brassica virus 3, broccoli mosaic virus, cabbage mosaic virus and cabbage virus B. This was first reported in *Brassica campestris* and *B. oleracea* from the U.S.A. by Tompkins (1937). The symptoms that persist among different species are vein-clearing or banding mosaic in *Arabidopsis thaliana*, *Brassica* spp., *Raphanus* spp. and other species of the Brassicaceae, Resedaceae.

Structure and Morphology

The virus consists of stable isometric particles, ~50 nm in diameter containing circular dsDNA encapsidated in subunits of a protein processed from a 58 kDa precursor. Major proteins include 37 K and 42 K proteins and consist of a central empty cavity of ~25 nm. It comprises 3 concentric shells, in which dsDNA is distributed in layers II and III along with the proteins. The central region is occupied by a solvent. Exposure to pH 11.25 leads to release of DNA tails without total disruption of the protein shell.

Genome Organization

Circular dsDNA molecules are of 8 kb with one discontinuity in one strand (minus strand or alpha strand) and two or three discontinuities in the (+)-strand. Discontinuities are associated with replication of the CaMVs. The DNA encodes 6 and possibly 8 genes. They are closely packed and the arrangement of the ORFs is shown in figure 3.11.

Proteins Encoded and their Function

There are 8 ORFs of which products of ORFs 3, 4 and 5 react with antiserum.

- ORF 1   ~36 K, play an important role in cell-to-cell movement (MP)

- ORF 2   ~19 K, involved in aphid transmissibility of virus, viroplasm
- ORF 3   14 K, non-sequence specific DNA binding protein, c-terminal part responsible for this activity, insect transmission of the virus
- ORF 4   57 K shell protein (coat), RNA binding activity
- ORF 5   79 K largest ORF in genome code for viral reverse transcriptase enzyme
- ORF 6   58 K viroplasm, role in disease induction, symptom expression
- ORF 7 and 8   may be involved in DNA binding activity

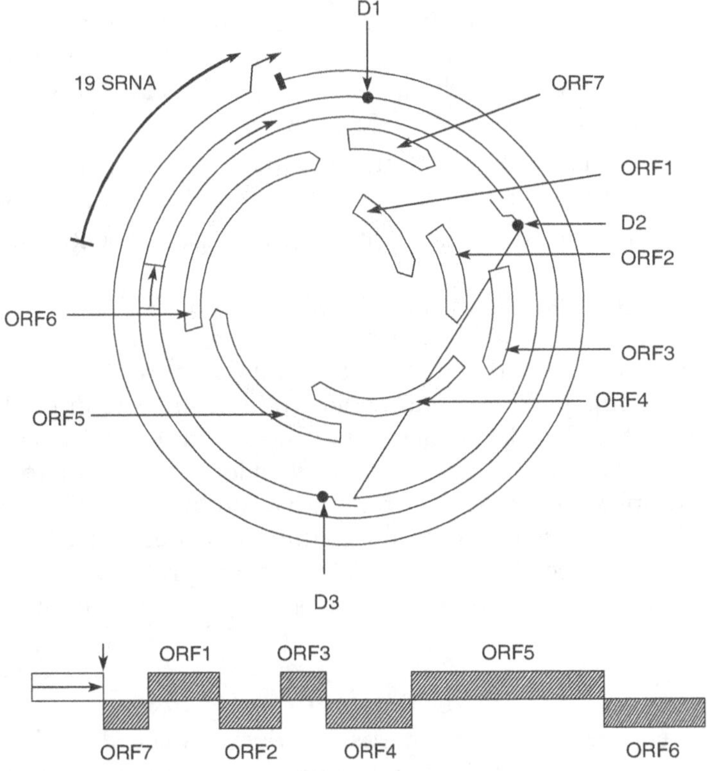

Figure 3.11   *CaMV genome organization.*

Expression of Genome

There are two phases in the nucleic acid replication cycle of CaMV: nuclear phase of transcription and cytoplasmic phase of gene expression and reverse transcription. The processes involved are:

a. dsDNA moves to the cell nucleus;

b. overlapping nucleotides at the gaps are removed and covalently closed to form complete dsDNA;

c. these minichromosomes form the template to transcribe 2 RNAs of 19S and 35S by host DNA-dependent RNA polymerase;

d. 35S promoter has 2 major domains: domain A (−90 to +8) root specific expression (numbering relative to transcription start site at +1); domain B (−343 to −90) expression in the aerial parts of the plant. ORF I–V are expressed from the 35S RNA;

e. domain B has 5 subunits B1 to B5, each conferring specific expression pattern in developing and mature leaves;

f. 19S promoter is monocistronic mRNA for ORF VI.

Long leader sequence of 600 nt and ORFs I–V downstream of putative ORF VII suggest with two unusual mechanisms of gene expression: "ribosome shunting" which involves the ribosome passing from a donor to acceptor site in the highly structured leader sequence (i.e., by-passing) and delivering the 40S ribosome subunit to the start codon of ORF I and form 80S complex "relay race" is transactivation by the product of ORF VI, after reaching the first termination codon of ORF, it does not leave the RNA, but reinitiates protein synthesis at the nearest AUG downstream or upstream from the termination site.

Replication

It encodes RT like retrovirus, but does not encode an integrase gene, therefore does not integrate into host genome for transcription, but does so from an episomal minichromosome.

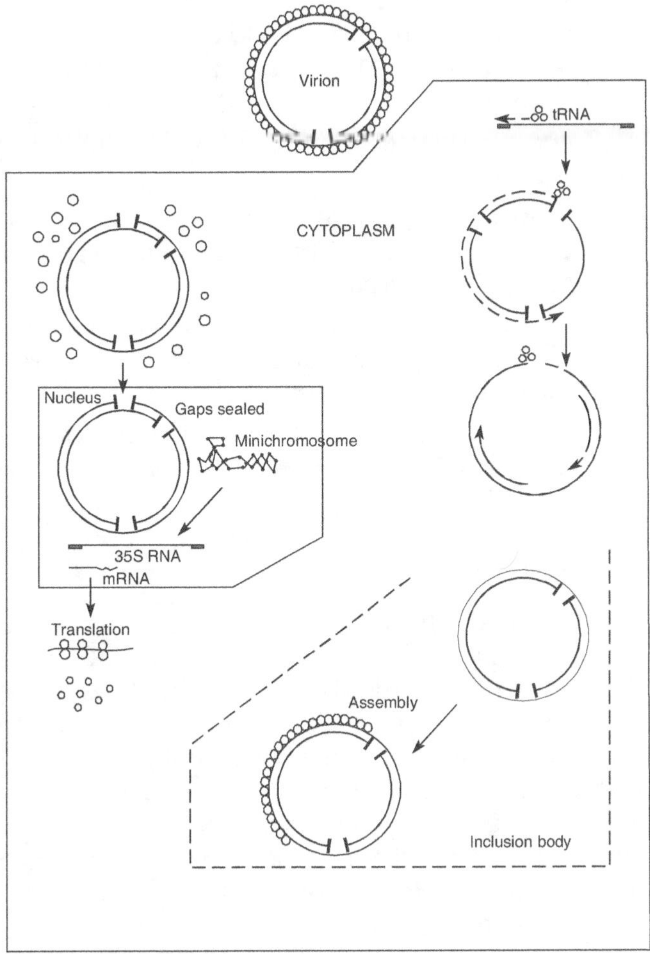

Figure 3.12   Replication of CaMV.

They are known as pararetroviruses. DNA in virus particle has discontinuities, but dsDNA in nucleus is supercoiled and associated with histones as a minichromosome. The pathway is outlined in figure 3.12.

- The 19S and 35S RNA transcribed in nucleus migrates to cytoplasm.

- Viral DNA synthesis on 35S RNA template is initiated by plant methionyl tRNA molecule at its 3´ downstream of D1 discontinuity in alpha strand DNA.

- Viral RT commences synthesis of (–)-strand of DNA and removes RNA from RNA–DNA hybrid giving a 'strong stop DNA'.

- Inclusion of bodies in cytoplasm include electron-dense type due to ORF III product and electron lucent type due to ORF II product.

- CaMV MP is a product of ORF I, a 38K protein (P1) that has nucleic acid binding activity, and passes through the plasmodesmata by tubules.

- CaMV induces chlorotic local lesions and a systemic mosaic in *Datura stramonuim*, *Nicotiana begerovii* and *N. edwardsonii*.

Transmission

Transmission of the virus is by vector insect—*Brevicoryne brassicae*, *Myzus persicae* and at least 25 other species of Aphididae—in a semi-persistent manner. Virus is lost by the vector when it moults because the virus does not multiply in the vector nor is transmitted congenitally to the progeny of the vector. It is transmitted only by mechanical inoculation. Helper component has to be acquired by an aphid either during or before the virus acquisition. Two non-capsid proteins of 18K (ORF II product P2/P18) and 15K (ORF III product P3) help in transmission.

The experimental host range includes almost less than three families that are susceptible. Experimentally infected plants mostly show vein clearing and mosaic symptoms. Diagnostically susceptible host species and symptoms include:

Vein clearing and also mosaic symptoms *in Brassica campestris* ssp. *napus, Brassica oleracea* var. *botrytis, B. campestris.*

Families containing susceptible hosts are Cruciferae (9/9), Resedaceae, Solanaceae (2/5) and families containing insusceptible hosts are Solanaceae (3/5).

Physical and Biochemical Properties

Properties of particles in sap   TIP: 75–80°C; LIV: 7–15 days. DEP: log10 minus 3. Infectivity of sap is not changed by treatment with diethyl ether. Leaf sap contains few virions. Electron microscopy: no special treatment required but virions flatten.

Cytopathology   Virions are found in all parts of the host plant in the cytoplasm and plasmodesmata. Inclusions are present in infected cells; and are of unusual shape with electron-dense matrices. Other cellular changes are deeply lobed nuclei, cell wall protrusions associated with vesicles and convoluted tubules.

COTTON LEAF VIRUS

Cotton leaf crumple virus (CLCrV) and cotton leaf curl virus (CLCuV) are two whitefly-transmitted geminiviruses that infect cotton. PCR products analysis indicated that the DNA A of CLCrV and CLCuV was approximately 2.6 kb and 2.7 kb, respectively. CLCrV and CLCuV are two distinct and distantly related geminiviruses. Recent changes in the sweet potato whitefly (*Bemisia tabaci* Genn.) populations have led to increased incidences of two important viral diseases of cotton (*Gossypium hirsutum* L.) caused by cotton leaf crumple virus (CLCrV) and cotton leaf curl virus (CLCuV). CLCrV was first reported from California and then Arizona a few years later. CLCrV symptoms are characterized by floral distortion, hypertrophy of interveinal tissue resulting in downward curling of leaves, and foliar mosaic accompanied by vein clearing and frequent vein distortion (figure 3.13 a, b). Losses resulting from CLCrV infection range from 21 to 86%, depending on the age of plants at the time of infection. CLCrV can be transmitted experimentally by *B. tabaci* to numerous species within the Malvaceae and Fabaceae families. Typical geminivirus particles were observed in partially purified CLCrV preparations.

(a)

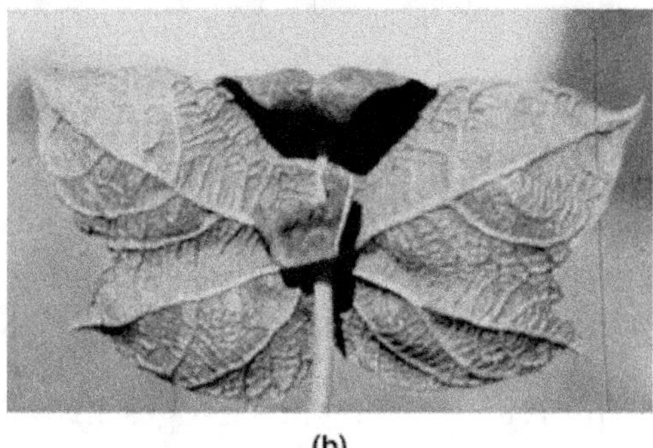

(b)

Figure 3.13   (a) Cotton plant infected with cotton leaf crumple virus (CLCrV) showing downward curling of leaves. (b) Foliar mosaic accompanied by vein clearing and frequent vein distortion.

CLCuV was first reported in Africa. Characteristic symptoms of CLCuV-infected cotton includes upward or downward curling of leaves, vein distortion and thickening, and enations on the underside of the leaves (figure 3.14 a, b). Infected plants bear few flowers and yield reduction varies with the age of plants at the initial infection. Severe epidemics of CLCuV have occurred in Pakistan in the past few years, with yield losses as high as 100% in fields where infection occurred early in the growing season. The virus can be experimentally transmitted by *B. tabaci* to cotton, *Nicotiana benthamiana*, cowpea, soybean, and okra.

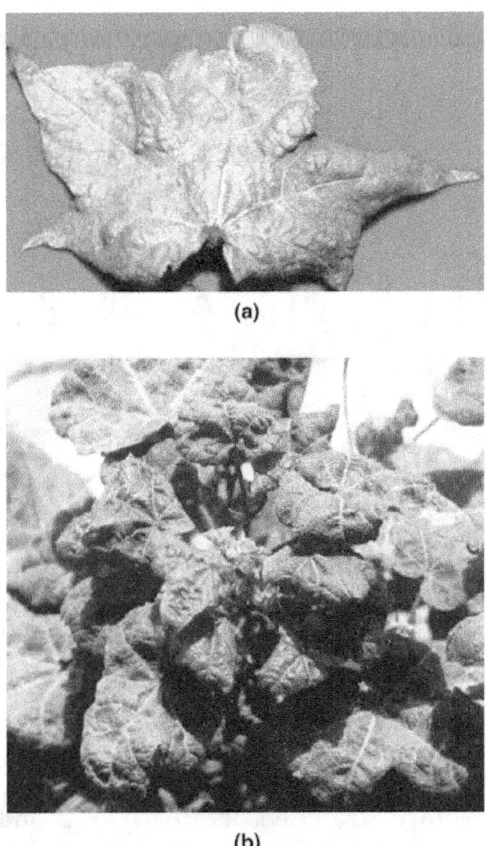

(a)

(b)

Figure 3.14 (a) Cotton plant infected with cotton leaf curl virus (CLCuV) showing curling of leaves. (b) Vein distortion and thickening, and enations on the underside of the leaves.

Geminiviridae consists of a diverse group of circular, single-stranded DNA viruses. Three genera of geminiviruses have been established on the basis of host ranges, vector relationships, and the number of genomic DNA components. Genus "Subgroup III Geminivirus" consists of all the whitefly-transmitted geminiviruses that infect dicotyledonous plants. With the exception of some tomato yellow leaf curl virus (TYLCV) isolates, the genome of this geminivirus subgroup

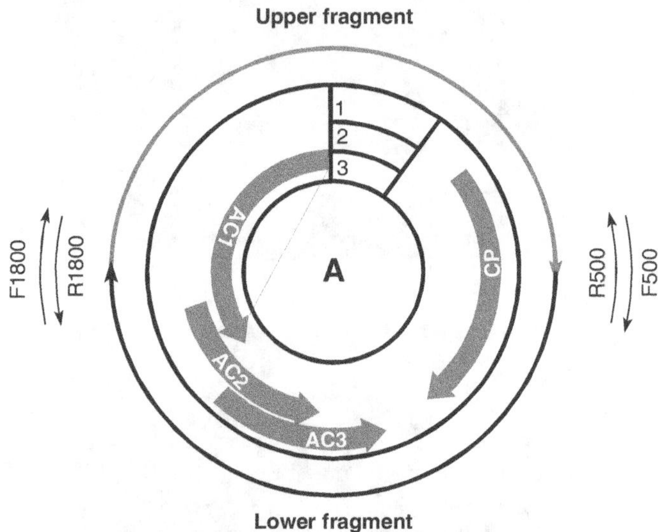

Figure 3.15   A strand of a subgroup III Geminivirus genome (bipartite) showing the transcriptional products.

consists of two DNA components, DNA A and DNA B. All the subgroup III geminiviruses with bipartite genomes sequenced to date have a similar genome organization. The virion strand of DNA A encodes the capsid protein gene while the complementary strand encodes AC1, AC2 and AC3 that are important for viral DNA replication and transcriptional regulation (figure 3.15). DNA B contains two genes that affect disease symptoms and viral movement in infected plants. There is a considerable degree of homology at the DNA level and the protein level within this geminivirus genus. Based on geographic distribution and variations in the genome organization, subgroup III geminiviruses can be further divided into the *Old World geminiviruses* and the *New World geminiviruses*. The former group occurs frequently in Africa and Asia and is represented by ACMV. The latter group occurs mainly in North and South

America and is represented by bean golden mosaic virus (BGMV) Both CLCrV and CLCuV infect dicotyledonous plants and are whitefly-transmitted. Previous studies suggested that they belong to the subgroup III geminiviruses. The sizes of DNA A component for CLCrV and CLCuV were estimated to be 2.6 and 2.7 kb, respectively. This size difference indicates that CLCrV and CLCuV are probably two different viruses. Reciprocal hybridization of PCR-amplified DNA fragments further suggests that CLCrV and CLCuV are two distinct but related viruses.

SUGARCANE MOSAIC POTYVIRUS

The synonyms includes grass mosaic virus, maize dwarf mosaic virus strain B (MacKenzie *et al.*, 1966; Louie and Knoke, 1975; Shukla *et al.*, 1989), sorghum red stripe virus. The Acronym is SCMV.

The different strains in this virus include the European maize dwarf virus, MDMV-B, SCMV-A, SCMV-B, SCMV-D, SCMV-E, SCMV-SC, SCMV-Sabi, sorghum concentric ringspot virus, possibly also abaca mosaic potyvirus.

This virus was first reported in *Saccharum* spp. by Brandes (1919); Gold and Martin (1955).

Natural Host Range and Symptoms

Symptoms persist.

- *Saccharum* spp., *Sorghum bicolor*, *Panicum* spp., *Eleusine* spp., *Setaria* spp., *Zea mays*—mosaic and/or ringspots, stunting.
- *Sorghum halepense* (particular strains)—mosaic and/or necrosis.

Transmission

Transmitted by a vector; an insect; *Dactynotus ambrosiae, Hysteroneura setariae, Rhopalosiphum maidis, Toxoptera graminum;*

Aphididae. Transmitted in a non-persistent manner. Virus is also transmitted by mechanical inoculation; by grafting; by seed; but not transmitted by pollen.

Diagnostically Susceptible Host Species and Symptoms

- *Saccharum* spp.—systemic mosaic.
- *Sorghum bicolor*—necrotic local lesion, then systemic mosaic, necrosis.
- *S. halepense*—systemic mosaic.
- *Zea mays*—systemic mosaic, ringspots.
- Most dicotyledonous plants.
- Maintenance and propagation hosts *Zea mays* (sweet corn).
- Assay hosts (Local lesions or Whole plants) *Sorghum* genotypes (L, W).

Susceptible Host Species

- *Anthoxanthum odoratum, Arundinaria amabilis, Arundo donax*
- *Cynodon dactylon, Digitaria decumbens, Echinochloa crus-galli*

Insusceptible Host Species

- *Agrostis tenuis, Avena sativa, Bromus mollis, Cortaderia selloana*
- *Dactylis glomerata, Festuca pratensis, Isachne globosa*

Families Containing Susceptible Hosts

- Gramineae (26/38), Leguminosae–Papilionoideae (1/1)

Families Containing Insusceptible Hosts

- Gramineae (12/38)

Particle Morphology

The virions are filamentous, not enveloped, usually flexuous, with a clear modal length of 730-755 nm and 13 nm wide. Axial canal is obscure and the basic helix is also obscure. The virions contain 5.5–6% nucleic acid; 94–94.5% protein. The genome consists of single-stranded RNA. Total genome size is 9.5 kb. Genome is unipartite; largest (or only) genome part is 9.5 kb. The virion protein(s) are one; M_r 34998 (328 residues (MDMV-B)).

Cytopathology

Virions are found in mesophyll, in cytoplasm and in Golgi apparatus (bundles, scrolls, tubes and laminated aggregates, namely inclusion group III). Inclusions are present in infected cells and are pinwheels; they do not contain virions. Other cellular changes include alteration of chloroplast organization (i.e., small grana with few thylakoids and uniform distribution of intergrana (stromal) lamellae.

Virus(es) with Serologically Related Virions

Using electro-blot immunoassay with cross-absorbed polyclonal antibodies against the N-terminus of the virion proteins, the potyviruses infecting sugarcane, maize and sorghum have been found to consist of four distinct types.

1. Johnson grass mosaic potyvirus—SCMV-JG and maize dwarf mosaic virus strain O.

2. Maize dwarf mosaic potyvirus—MDMV-A, MDMV-D, MDMV-E and MDMV-F.

3. Sugarcane mosaic potyvirus—MDMV-B, SCMV-A, SCMV-B, SCMV-D, SCMV-E, SCMV-Sc, SCMV-Bc and SCMV-Sabi.

4. Sorghum mosaic potyvirus—SCMV-H, SCMV-I and SCMV-M (Shukla *et al.*, 1989).

COTTON LEAF CRUMPLE BIGEMINIVIRUS

Cotton leaf crumple virus CLGV was first reported in *Gossypium hirsutum*, from California, U. S. A. Mild and severe strains are usually recognized.

Natural Host Range and Symptoms

Symptoms persist.

- Species of *Gossypium, Abutilon, Althaea, Hibiscus, Malva* and of *Castanospermum, Glycine, Phaseolus, Vicia—* mosaic and malformation.

Transmission

The virus is transmitted by a vector; an insect called as *Bemisia tabaci* belonging to the family Aleyrodidae. They are transmitted in a semi-persistent manner. The virus gets lost by the vector when it moults and it does not multiply in the vector. It is not transmitted congenitally to the progeny of the vector or by mechanical inoculation; is transmitted by grafting; It is neither transmitted by contact between plants nor by seed or pollen.

The host range includes about less than three families that are susceptible. Experimentally infected plants mostly show chlorosis and malformation.

Diagnostically Susceptible Host Species and Symptoms

These species include: *Malva parviflora, Gossypium hirsutum* 'Delta Pine 70´ , *Vigna angularis* (Azuki bean) and *Castanospermum australe* (Delgado bean)—systemic chlorosis and malformation.

Diagnostically Insusceptible Host Species

These species include: *Cucumis melo, Abelmoschus esculentus, Spinacia oleracea, Nicotiana glutinosa,* and *Zinnia elegans.*

Maintenance and Propagation Hosts

These include *Gossypium hirsutum*, *Malva parviflora* and a useful purification host *Phaseolus vulgaris*.

Assay hosts (Local lesions or Whole plants)

The assay hosts include *Gossypium hirsutum* (W) and *Malva parviflora* (W).

Susceptible Host Species

These species include

- *Abutilon, Althaea, Castanospermum, Castanospermum australe*
- *Glycine, Glycine max, Gossypium, Gossypium hirsutum*
- *Hibiscus, Hibiscus cannabinus, Malva, Malva parviflora*

Insusceptible Host Species

These include

- *Abelmoschus esculentus, Cucumis melo, Nicotiana glutinosa*
- *Spinacia oleracea, Zinnia elegans*

Families Containing
Susceptible and Insusceptible Hosts

Susceptible:

- Leguminosae—Papilionoideae (9/9), Malvaceae (8/9)
  Insusceptible:
- Chenopodiaceae (1/1), Compositae (1/1), Cucurbitaceae (1/1)
- Malvaceae (1/9), Solanaceae (1/1)

Particle Morphology

Virions are geminate. They are not enveloped, and are 17–20 nm in diameter; dimers are of 30–32 nm in length.

Cytopathology

Virions are found in phloem. Inclusions present in infected cells (in phloem and parenchyma only) are unusual in shape, spherical or amorphous. They are numerous in younger leaves. Other cellular changes include hypertrophy of cells containing inclusions.

SUGARCANE STREAK MONOGEMINIVIRUS

Sugarcane streak monogeminivirus consists of two strains namely, Natal type and Mauritus.

The virus was first reported in *Saccharum officinarum*, from Natal, South Africa, by Storey (1925).

Natural Host Range and Symptoms

Symptoms persist. Symptoms include chlorotic spots and narrow chlorotic streaks.

Narrow, elongated chlorotic spots and streaks are found in *Cenchrus echinatus, Saccharum officinarum*.

Transmission

The virus is transmitted by a vector, an insect, *Cicadulina mobila, C. bipunctata bipunctata. C. bipunctella zeae*; Cicadellidae. Principal natural vector(s) includes: *Cicadulina mbila, C. bipunctata bipunctata*, and *C. bipunctella zeae*. They are transmitted either in a non-persistent or a semi-persistent manner. Virus gets retained when the vector moults and does not multiply in the vector; not transmitted congenitally to the progeny of the vector; does not require a helper virus for vector transmission; not transmitted by mechanical inoculation; not transmitted by grafting; not transmitted by contact between plants; not transmitted by seed and not transmitted by pollen.

Experimental Host Range

Several (3–9) families are susceptible. Experimentally infected plants mostly show mild chlorotic streaks and spots.

Diagnostically Susceptible and Insusceptible Host Species and Symptoms

Susceptible

- *Saccharum officinarum*—yellow and white streaks
- *Cenchrus echinatus*
- *Eleusine indica*—mild permanent leaf streaks
- *Zea mays*—mild, short, pale streaks

Insusceptible

- *Digitaria horizontalis.*

Particle Morphology

Virions are geminate and not enveloped. They are about 20 nm in diameter and 30 nm in length.

Biochemical Properties

a. Virions consist of 0% lipid.
b. The genome consists of DNA which is single-stranded and circular. Total genome size is 2.76 kb.
c. The genome is unipartite; the largest (or only) genome part is 2.76 kb. Base composition is 27.7% G; 21.8% A; 23.1% C and 28.1% T.

Features of the Genome

The features of the genome include 2 virion sense ORFs encoding proteins of 11,000 and 27,000 M_r; the smaller ORF has homology with V1 of MSV (transport protein), and the larger with V2 of MSV (capsid protein). There are two ORFs in the complementary strand; C1 encoding a protein of 34,940 M_r and C2 of 17,000 M_r, these, like other monogeminiviruses, may be spliced to give c′, which encodes the replicase protein of 41,000 M_r. There is a TATA box upstream of V1 and upstream of C1, and an AATAA consensus polyadenylation signal downstream of C2 and V2. Sequence with homology to the core upstream activator sequence of MSV is upstream of C1. There are two intergenic regions. The large one is between the starts of C1 and V1 and contains promoter sequences and a conserved stem and loop structure with the TAATATTAC sequence common to all geminivirus genomes. The small intergenic region lies between the termini of V2 and C2 and contains AATAA sequences and a possible binding site of about 72 nucleotides in the genomic strand for an endogenous DNA primer molecule. Non-genomic nucleic acid is not found in the virions. The features of proteins include: virion protein(s) one; M_r 28000; capsid protein. This genome replicates in nuclei. Replication does not depend on a helper virus.

Cytopathology

Virions are found in leaves, mesophyll, meristem and phloem nuclei. Inclusions present in infected cells are unusual in shape and non-crystalline.

SUGARCANE BACILLIFORM BADNAVIRUS

Sugar cane bacilliform badnavirus SCBV was first reported in *Saccharum officinarum*. No conspicuous symptoms were found in this host.

Diagnostically Susceptible and Insusceptible Host Species and Symptoms

Susceptible

- *Saccharum officinarum* (clone CP 44-101)—symptomless systemic infection.

Insusceptible

- *Chenopodium quinoa*
- *Nicotiana benthamiana*
- *N. glutinosa*
- *Sorghum halepenses*
- *Zea mays*

Particle Morphology

Virions bacilliform; not enveloped; 131 nm in length; 31 nm wide. These viruses belong to the Badnavirus taxonomy. The virus(es) with serologically related virions include banana streak badnavirus.

TOMATO YELLOW MOSAIC VIRUS

Tomato yellow mosaic virus has strains like the tomato yellow leaf curl and tomato yellow mosaic viruses. They induce similar symptoms but do not seem to be related. It was first reported in *Lycopersicon esculentum*.

Natural Host Range and Symptoms

Symptoms persist.

- *Lycopersicon hirsutum*—yellow mosaic
- *L. esculentum*—yellow mosaic, stunting, small leaves
- *L. pimpinellifolium, Solanum tuberosum*—mild yellow mosaic.

Transmission

Transmitted by a vector, an insect, *Bemisia tabaci*; Aleyrodidae. Transmitted in a persistent manner (for up to 7 days). Virus gets retained when the vector moults; does not multiply in the vector. It is not transmitted congenitally to the progeny of the vector. It is transmitted by mechanical inoculation (poorly when inoculum is from an old tissue) and by grafting. It is not transmitted by contact between plants nor by seed and pollen.

Diagnostically Susceptible Host Species and Symptoms

Susceptible

- *Datura stramonium, Nicotiana benthamiana*—systemic yellow mosaic.
- *Nicandra physalodes, Petunia hybrida, Physalis peruviana*—systemic mild yellow mosaic.
- *Nicotiana glutinosa*—systemic yellow spots at the base of first infected leaves, then mosaic.
- *N. tabacum* cv. Samsun—systemic mosaic and leaf deformation.

Insusceptible

- *Chenopodium quinoa*
- *Euphorbia heterophylla*
- *Phaseolus vulgaris*
- *Vigna unguiculata*
- *Sida rhombifolia*
- *Datura metel*

Families Containing Susceptible and Insusceptible Hosts

Susceptible

- Solanaceae (13/16)

Insusceptible

- Chenopodiaceae (3/3)
- Cruciferae (2/2)
- Cucurbitaceae (3/3)

- Euphorbiaceae (1/1)
- Leguminosae–Papilionoideae (6/6)
- Malvaceae (2/2)
- Solanaceae (3/16)

Particle Morphology

Virions geminate; not enveloped; 18–20 nm in diameter; angular in profile; without a conspicuous capsomere arrangement (figure 3.16).

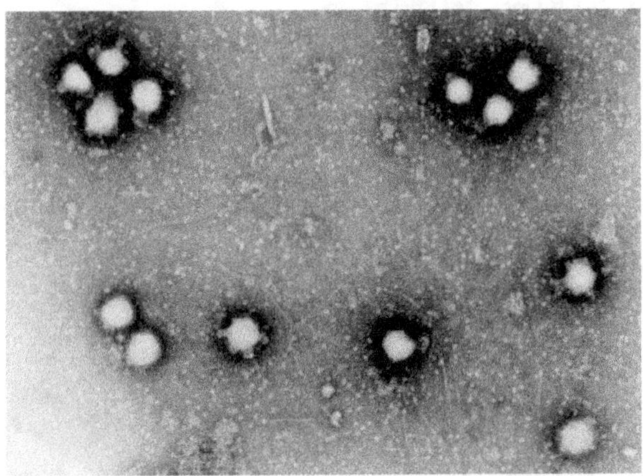

Figure 3.16   Electron micrograph of tomato yellow mosaic virus (TYMV).

Physical and Biochemical Properties

One sedimenting component in purified preparations; sedimentation coefficient is 70S. The virion contains 20% nucleic acid; 80% protein; 0% lipid. It also consists of DNA which is single-stranded and circular. Total genome size is 5.096 kb. Genome consists of two parts; largest (or only) genome part is 2.588 kb; the 2nd largest is 2.508 kb. Infectivity gets retained when deproteinized with phenol or detergent. Additional factor not required for infectivity. Non-genomic nucleic acids are not found in the virions. The virion protein(s) one; M_r 28000 is a coat protein. The genome replicates in nuclei. Replication does not depend on a helper virus.

Cytopathology

Virions are found in phloem nuclei and nucleoli. Inclusions present in infected cells are unusual in shape; clusters of virions. Other cellular changes include fibrillar rings in the nucleus that may be where virions assemble, and are composed of DNA and protein.

RICE STRIPE VIRUS

This is a virus with filamentous particles, 500–2000 nm long, but only about 8 nm wide (figure 3.17). The particles sediment as four components and contain four ssRNA species, a major protein (the nucleocapsid protein) and a minor protein (the RNA polymerase). RNA-1 is negative-stranded and RNAs 2 to 4 are ambisense. Non-structural protein material is produced abundantly in infected cells. Mechanical inoculation is difficult.

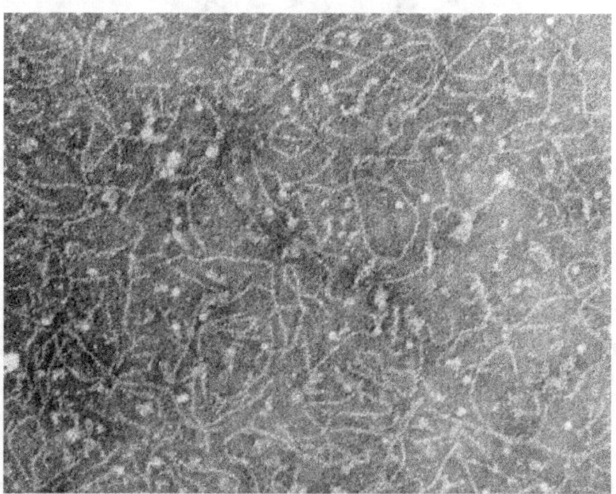

Figure 3.17   Rice stripe virus (RSV) showing filamentous morphology with a range of 500–2000 nm length and about 8 nm width.

The virus is transmitted by *Laodelphax striatellus* and three other planthopper species in a persistent manner; it is transmitted through the egg to about 90% of progeny insects. It infects many species of Gramineae. It occurs in rice-growing areas of Asia and the USSR, and causes significant reduction in rice yield.

Main Diseases

In rice the virus causes chlorotic stripes, chlorosis, moderate stunting and loss of vigour (figure 3.18). In severe infections the leaves develop brown to grey necrotic streaks and die.

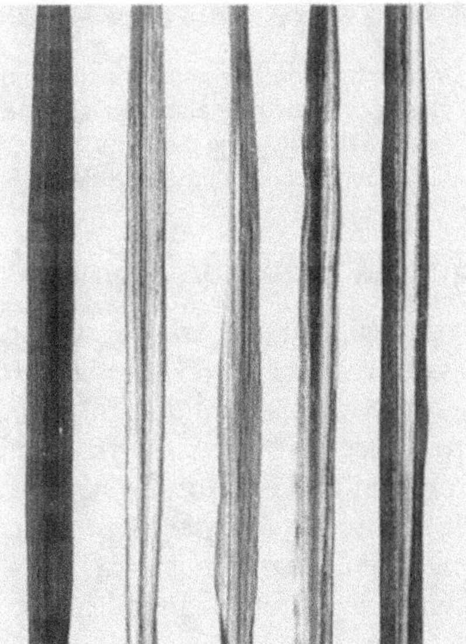

Figure 3.18 Chlorotic stripe symptoms on infected leaves of rice *(Oryza sativa)* compared with a healthy leaf (extreme left).

Diseased plants produce few or no panicles: those produced carry whitish to brown discoloured malformed spikelets (figure 3.19). Early infection of rice causes significant loss of yield; late infection also reduces yield by retarding ear emergence and ripening. In maize and wheat the virus causes chlorotic stripes, chlorosis and stunting.

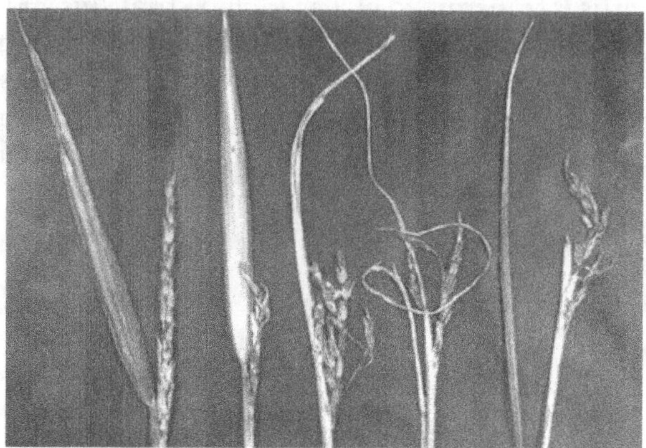

Figure 3.19　Malformed and immature ears, and chlorotic stripes on flag-leaves, of rice plants infected at a late growth stage, seen with few or no panicles that carry whitish to brown discoloured malformed spikelets.

Host Range and Symptomatology

RSV occurs naturally in rice, maize, wheat, oat, foxtail millet and wild grasses such as *Cynodon dactylon, Digitaria adscendens, D. violascens, Eragrostis multicaulis* and *Setaria viridis.* The virus is reported to infect 37 species of the family Gramineae. The main symptom induced is chlorotic striping. Non-gramineous plant species, such as *Cyperus amuricus* var. *laxus, C. sanguinolentus,* and *Eriocarlan robustius* are also reported as host plants. The virus is readily transmitted to test plants by the planthopper, *Laodelphax striatellus,* but is difficult to transmit mechanically.

Diagnostic Species

Oryza sativa (rice)

Seedlings of most Japanese paddy varieties are highly susceptible to infection. Chlorotic stripes, mostly light yellowish broken streaks, develop on systemically infected leaves 10 to 25 days after inoculation by planthoppers

(L. striatellus). Characteristically, limp chlorotic leaves emerge without unfolding, elongate, droop and wilt (figure 3.20). Yellowing and moderate stunting also occur.

Triticum aestivum (wheat) and *Zea mays* (maize) (cv. golden cross bantam)

Symptoms in these species are similar to those in *O. sativa*. Wheat plants may produce whitish, rolled, needle-like leaves that quickly droop.

Figure 3.20 Characteristic drooping and wilting symptoms in young seedlings of *O. sativa* due to RSV infection.

Transmission by Vectors

The most active vector in the field is the small brown planthopper, *Laodelphax striatellus*; three other planthopper species, *Unkanodes sapporona*, *U. albifascia* and *Terthron albovittatum* also transmit. The proportion of active transmitters of *L. striatellus* is about 20%. Although the shortest acquisition feeding period is 15 min, best transmission is obtained with planthoppers that acquire the virus by feeding for 1 day. The incubation period of the virus in *L. striatellus* ranges from 5 to 21 days but incubation is complete within 5 to 10 days for most individuals. Minimum inoculation feeding time is 3 min; about half or more of the infective planthoppers infect rice seedlings after feeding for 1 h. Ability of the insects to transmit the virus decreases markedly with age. Females of *L. striatellus* are more efficient transmitters than males. The virus passes through a high percentage of eggs to progeny, about 90% in *L. striatellus*. *L. striatellus* was selected and bred for high or low ability to acquire and transmit the virus (50–60% and less than 10% of the insects transmitting, respectively), and this was correlated with the frequency of transmission of the virus through the eggs. Evidence of virus multiplication in *L. striatellus* has been obtained by repeated passage through the eggs for 6 years through 40 generations, by serial transfer of the virus from insect to insect by injection and by detection of virus particle antigen in various organs of viruliferous insects by fluorescent antibody staining.

Serology

RSV is a good immunogen. Rabbit antisera with titers of 1/512 to 1/1024 in precipitin tests are readily obtained. In agar gel diffusion tests, it is necessary to disrupt virus particles in purified preparations or crude sap of infected tissues by treatment with 0.5% SDS. The haemagglutination test was used to detect the virus particle antigen at high dilutions in vectors or plants. Monoclonal antibodies have been made and used for specific detection of the viral antigen and ELISA is useful for detecting the virus in plants and for practical applications, the latex flocculation test, rapid immunofilter paper assay and dot immunobinding assay (DIBA) are useful.

Particle Composition

Nucleic acid   The nucleic acid found in predominance is ssRNA. However, dsRNAs containing the sequence of viral ssRNAs are also detected in agarose gel electrophoresis. The two smallest RNAs, RNA-3 and RNA-4, occur in M component (which, as mentioned above, is a mixture of two components, M1 and M2), and RNA-2 occurs in B component. The nB component contains RNAs 2 to 4 together with the largest RNA,

RNA-1, and only this nB component is infectious (figure 3.21). The RNA constitutes about 12% of the particle weight, as judged by the A_{260}/A_{280} value of purified preparations. The three RNA species from M and B components are linear molecules about 0.7–1.0 μm long.

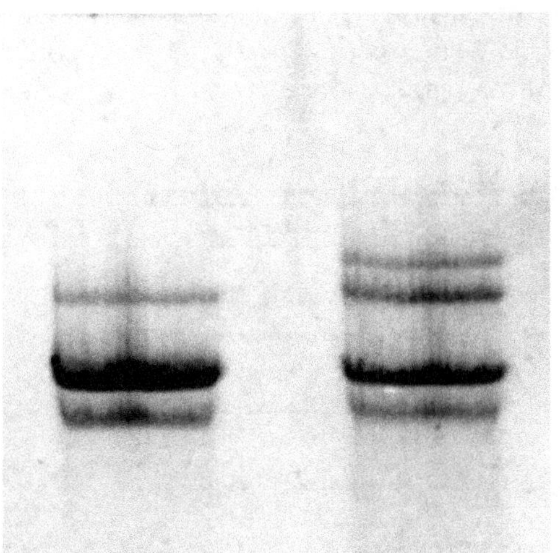

Figure 3.21   Ribonucleic acid of RSV on polyacrylamide gels: *(left),* three RNA segments from M+B components; *(right),* four RNA segments in a preparation from nB.

Protein   One species of nucleocapsid protein, of 32 kDa. A minor, but large molecular weight protein, RNA-dependent RNA polymerase, is also associated with viral particles.

Genome Properties

The sequence of about 20 bases at the 5´- and 3´-termini of each RNA are almost complementary to each other; ten nucleotides of these termini are identical among RNAs 1 to 4, except for one base change in RNA-1.

The genome organization is shown in figure 3.22. RNA-1, which encodes the RNA-dependent RNA polymerase, is negative-stranded and RNAs 2 to 4 are ambisense. In each of RNAs 2 to 4, the intergenic non-coding region between the two ORFs contains an oligo(A)- and oligo(U)-rich sequence that can be formed into base-paired hairpin stems. The nucleocapsid protein is encoded at the 5´-proximal region of virus complementary-sense RNA-3, and the major non-structural protein is encoded at the 5´-proximal region of virus - sense RNA-3. The major non-structural protein is synthesized in an in vitro translation system by using the viral RNA, but

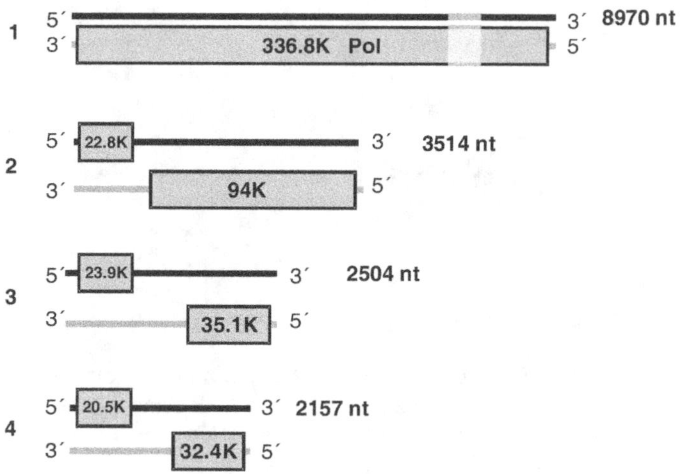

Figure 3.22 Genome organization of RSV (isolate T). The ORFs of four RNA segments are displayed on viral-sense RNAs as a black bar and on viral complementary- sense RNAs as a blue bar and on viral complementary-sense RNAs as a grey bar, the proteins deduced from each ORF are shown as boxes.

nucleocapsid protein is not synthesized. mRNAs with the genomic size of these proteins are produced in vivo, and they have an additional 10 to 23 non-viral nucleotides at the 5´-terminus. At present there are no experimental data on the functions of the ORFs on viral-sense RNAs 2 and 3, or on viral complementary-sense RNA-4.

POTATO VIRUS X (PVX)

Morphology

PVX infects plants and fungi. Virions are not enveloped, nucleocapsids filamentous; usually flexuous; with a clear modal length 445–523.1–775 nm long; 11–12.8–15 nm in diameter (figure 3.23). Symmetry is helical. Axial canal is obvious or obscure . Axial canal is 3.4–6.3–12 nm in diameter. Basic helix is obvious or obscure. Pitch of helix is 2.8-3.331-3.5 nm.

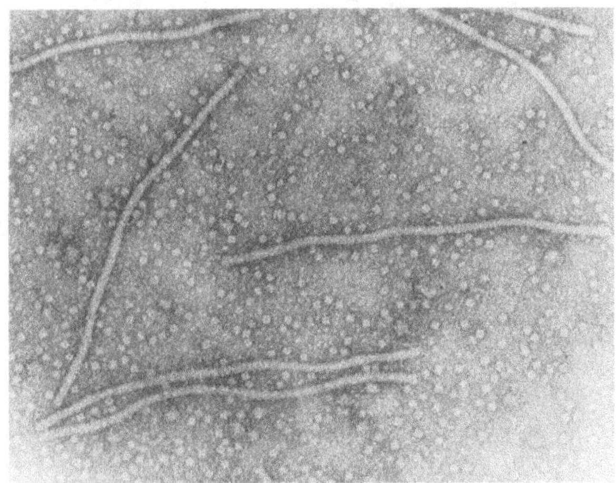

Figure 3.23   Electron micrograph of Potato virus X (PVX) showing filamentous morphology with a clear modal length 445-523.1-775 nm long; 11-12.8-15 nm in diameter.

Physicochemical and Physical Properties

Buoyant density is about 1.28–1.308–1.33 g cm^{-3} in CsCl. One sedimenting component in purified preparations, or two sedimenting components in purified preparations. Sedimentation coefficient is 110–122–144 S; of the other(s) 102–111.6–122 S. Isoelectric point is at a pH 4–4.75–5.3. A260/A280 ratio 1.18-1.229-1.4. Infectivity of sap is not changed by treatment with diethyl ether. Infectivity gets retained when deproteinized with proteases; gets retained when deproteinized with phenol or detergent.

Nucleic Acid

Virions contain 5–5.755–8% of nucleic acid. They contain one molecule of linear positive-sense single-stranded RNA. Total genome length is 5845–6725–8100 nt. The largest segment is 5845–6725–8100 nt. Base ratio consists of 15.5–21.12–25% guanine; 26.4–29.9–33.8 % adenine; 23.4–26.02–30.3% cytosine and; 20.5–22.92–25.7% uracil. The 5´ end of the genome has a cap, or an unusual type. Cap sequence is m7GpppA. Poly(A) region is present (5/5). 3´ end has a poly(A) tract. Non-genomic nucleic acid are either found or not found in the virions. Sub-genomic YRNA are found in infected cells. Four virus-specified dsRNA species are found in infected cells. Size of the largest virus specified dsRNA is 6–11.55–17.1 kbp, 2nd largest 4.2–5.6–7 kbp. 3rd largest is 1.7-3.7-5.7 kbp, and the 4th largest is 0.5-2.55-4.6 kbp.

Proteins

Virions contain 92–94.14–95% protein. One structural or two structural virion proteins are found. Protein size is about 14000-24340-30000 Da. Protein size of the 2nd largest is about 30500 Da. Virion structural proteins are not glycosylated. Virus-coded non-structural proteins are isolated, or identified by genomic sequence analysis. One or five non-structural proteins are found. Protein size is about 115900-141800-159900 Da. Protein size of the 2nd largest is about 25340-45670-66000 Da, 3rd largest is 11570-27780-44000 Da, 4th largest is 11210 Da, and the 5th and smallest is 8079 Da.

Lipids

Virions contain 0% lipid.

Genome Organization and Replication

Genomic nucleic acid is infectious.

Function of helper and satellite viruses Virions are associated with helper virus, but independent from its functions during replication.

Cytopathology Virions are also found in cytoplasm, or in nuclei, or in cell vacuoles. Inclusions are present in infected cells, or absent from infected cells. Inclusions are crystals in the cytoplasm, crystals in the nucleus, amorphous X-bodies, or viroplasma, or may be unusual in shape. They contain virions, or they do not contain virions.

Natural Host Range and Symptoms

Symptoms persist (23/29), or vary seasonally (3/29), or none (4/29).

Experimental Host Range and Symptoms

Many (>9) families are susceptible (4/40), or several (3–9) families are susceptible (23/40), or few (<3) families are susceptible (13/40).

Susceptible Hosts

Experimentally infected species susceptible to virus:

a. *Agapanthus praecox* ssp. *orientalis* (1/52)
b. *Allium cepa* (2/52)
c. *Allium cepa* var. *ascalonicum* (2/52)
d. *Allium chinense* (1/52)
e. *Allium porrum* (1/52)
f. *Allium vineale* (2/52)
g. *Amaranthus caudatus* (5/52)
h. *Amaranthus hybridus* (1/52)
i. *Amaranthus retroflexus* (3/52)
j. *Anthriscus cerefolium* (1/52)
k. *Antirrhinum majus* (4/52)

    l.   *Apium graveolens* (2/52)

    m. *Arachis hypogaea* (2/52), etc.

Insusceptible Hosts

Species inoculated with virus, but tested not to be susceptible:

    a. *Abelmoschus esculentus* (1/34)

    b. *Ageratum conyzoides* (1/34)

    c. *Allium cepa* (3/34)

    d. *Allium fistulosum* (3/34)

    e. *Allium porrum* (1/34)

    f. *Allium sativum* (1/34)

    g. *Allium schoenoprasum* (1/34)

    h. *Allium tuberosum* (1/34)

    i. *Amaranthus bicolor* (1/34), etc.

Families containing susceptible hosts

    a. Alliaceae (3/52)

    b. Amaranthaceae (28/52)

    c. Amaryllidaceae (2/52)

    d. Apocynaceae (4/52)

    e. Araceae (1/52)

    f. Asparagaceae (1/52)

    g. Balsaminaceae (1/52)

    h. Basellaceae (1/52)

    i. Cactaceae (3/52), etc.

Families containing insusceptible hosts

    a. Alliaceae (5/34)

    b. Amaranthaceae (10/34)

    c. Amaryllidaceae (2/34)

    d. Anthericaceae (2/34)

    e. Apocynaceae (3/34)

f.   Araceae (1/34)

g.   Caricaceae (1/34), etc.

Transmission

Transmitted by either involving (11/22), or not involving a vector (11/22). Virus transmitted by mechanical inoculation (43/46), or not transmitted by mechanical inoculation (3/46); transmitted by grafting (7/7); transmitted by contact between plants (5/11), or not transmitted by contact between plants (6/11); transmitted by seed (2/19), or not transmitted by seed (17/19); not transmitted by pollen (4/4). Transmitted by an insect (8/10), or a mite (2/10); Aphididae (7/8), or an unusual insect group (1/8); Eriophyidae (2/2). Transmitted in a non-persistent manner (6/6). Virus requires, for vector transmission, a helper virus (2/3), or does not require a helper virus for vector transmission (1/3).

Geographic Distribution

Probably distributed worldwide (8/8).

Diagnostic Methods

Leaf sap contains few virions (10/31), or contains many virions (21/31).

Taxonomic Structure

Type species   potato virus X.

Species   Asparagus virus 3; cactus virus X (58); cassava virus X; clover yellow mosaic virus (111); commelina virus X; cymbidium mosaic virus (27); foxtail mosaic virus (264); hydrangea ringspot virus (114); lily virus X; narcissus mosaic virus (45); nerine virus X; papaya mosaic virus (56); pepino mosaic virus; plantago severe mottle virus; plantain virus X (266); tulip virus X (276); viola mottle virus (247); white clover mosaic virus (41).

Tentative species Artichoke curly dwarf virus; bamboo mosaic virus; barley virus B1; boletus virus; cassava common mosaic virus (90); centrosema mosaic virus; daphne virus X (195); dioscorea latent virus; lychnis potexvirus; malva veinal necrosis virus; nandina mosaic virus; negro coffee mosaic virus; parsley virus 5; parsnip virus 3; parsnip virus 5 and so on.

REVIEW QUESTIONS

1. Write short notes on:
   a. Icosahedron
   b. Three examples of viruses of algae
   c. Cryptoviruses
   d. Aphids
   e. Plasmodesmata
   f. Bimodal transmission
   g. Movement proteins (MP)
   h. Inclusion bodies
   i. Thrips
   j. Viroids
   k. Satellite RNAs
   l. PVX
   m. CaMV
   n. Indicator plants

2. Explain in detail the morphology, replication and life cycle of TMV.
3. What are the modes of transmission of plant viruses?
4. What are the common disease symptoms of plant diseases caused by viruses?
5. Give an account of the diagnostic techniques of plant viral infections.

6. Explain the non-persistent, semi-persistent and persistent modes of transmission of viruses.

7. Give an account of the control of plant viruses.

8. Give a note on potyviruses.

9. Explain the structural composition, replication and transmission of cauliflower mosaic virus (CaMV).

10. Give a note on cotton leaf virus.

4   ANIMAL VIROLOGY

This chapter deals with the study of viral diseases and other infections caused by viruses among 'animals' (both human beings and animals).

POXVIRUSES

Poxviruses are the largest and most complex of all viruses; the subfamily chordopoxvirinae (poxviruses of vertebrates) contains eight genera of which members of genera *Orthopoxvirus*, *Parapoxvirus*, *Yatapoxvirus* and *Molluscipoxvirus* infect humans. The poxviruses are neither icosahedral nor helical; their structure is brick-shaped with rounded corners, 250 × 200 × 200 nm, irregular arrangement of tubules on outer membrane, which is sometimes surrounded by an envelope, referred to as complex (figure 4.1). In parapoxvirus, the virions are ovoid, 260 × 160 nm, with regular spiral arrangement of tubule on

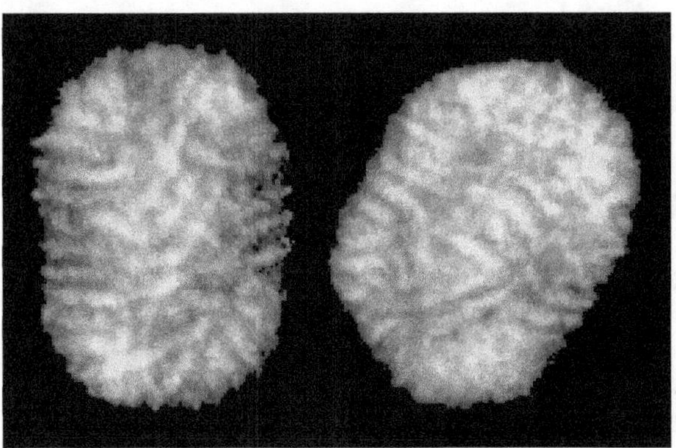

Figure 4.1   Orthopox virus.

outer membrane. Inside there is a core structure shaped like a dumbbell, and two accompanying lateral bodies, so named after their location in the virion (figure 4.2).

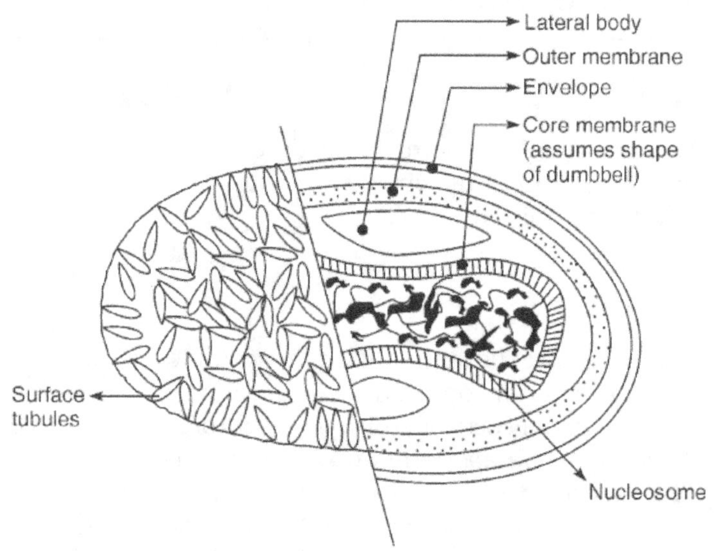

Figure 4.2   Diagram of the structure of the virion of vaccinia virus.

Genome

The nucleic acid is a linear dsDNA, 130 kbp (Parapoxvirus); 145 kbp (Yatapoxvirus), and 170–25 kbp (Orthopoxvirus) to 220 kbp (Cowpox). It codes for more than 100 polypeptides, including a DNA-dependent RNA polymerase and other enzymes.

Replication

The poxviruses replicate only in the cytoplasm, and enveloped particles are released by exocytosis, while non-enveloped particles are released by cell lysis. The virion enters the cell either by endocytosis or by a fusion event. The viral core enters the cytoplasm and acts as a scaffolding for subsequent replication events (figure 4.3). Furthermore, the virus transports many essential enzymes such as viral transcriptase, transcription factors, capping and methylating enzymes and a poly(A) polymerase. Transcription of viral DNA is initiated rapidly and ~100 early viral genes are activated, particularly genes coding for enzymes involved in viral DNA replication. Transcription of intermediate and late genes is initiated at the same time as DNA replication. Virus assembly takes place in particular areas of the cytoplasm, and immature viruses can be visualized quite easily. During maturation virions move to the golgi complex where they are enveloped before being released by budding or by cell disruption. Enveloped virions may have some advantages, including speed of uptake by cells.

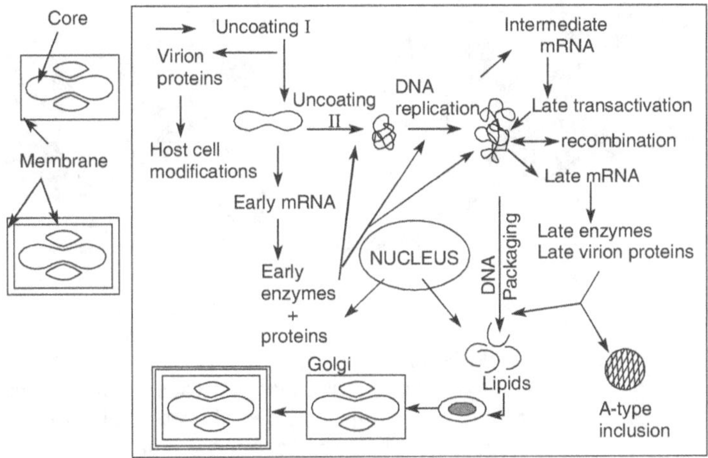

Figure 4.3   Diagram showing poxvirus replication.

Clinical Aspects

There were two main categories of small pox, caused by slightly different viruses. Variola major caused a mortality of around 30% whereas variola minor, or alastrim, killed less than 1% of its victims. The incubation period was usually 10–12 days; a febrile illness of sudden onset lasting 3–4 days was followed by the appearance of a rash progressing from macules to papules, vesicles, and pustules which then formed crusts. Surviving patients were often left with unsightly scars or pockmarks. The distribution of the rash was centrifugal (figure 4.4) i.e., it affected the extremities more than the trunk, as opposed to the centripetal rash of chickenpox.

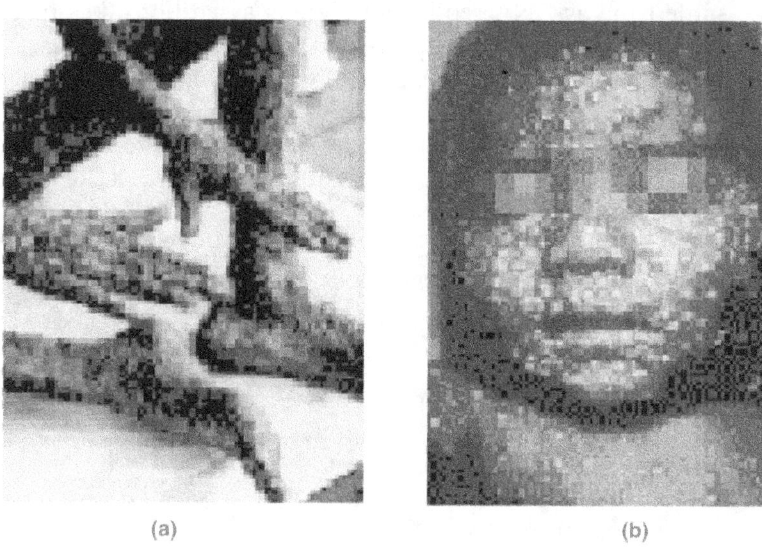

Figure 4.4 Variola lesion: (a) Characteristic skin lesion of variola on the arms and legs of an adolescent. (b) Unvaccinated infant with centrifugally distributed umbilicated pustules on day 7 of ordinary form of variola major strains of smallpox.

Laboratory Diagnosis

The differential diagnosis most often needed was between smallpox and chickenpox, which could on occasion resemble each other clinically. Electron microscopy of vesicle fluid readily distinguishes between pox and herpes virions. Most poxviruses can be propagated on the chick embryo chorioallantoic membrane, on which they form circumscribed pocks, 2–3 mm in diameter, or in cell cultures. Molluscum has so far not been grown in the laboratory.

Control

Immunization by smallpox vaccine.

HUMAN ORTHOPOXVIRUSES

Smallpox

In 1967 the World Health Organization (WHO) embarked on smallpox eradication campaign. The world's last naturally occurring case of small pox was recorded in October, 1977. On May 8, 1980, the WHO formally announced the global eradication of small pox. Eradication of small pox could be achieved because of

- its low communicability,
- no subclinical infection or carrier state,
- a single serotype virus which induced good immunity,
- an effective vaccine (originally discovered by Jenner in 1796),
- no animal reservoir and
- aggressive surveillance-containment measures.

Routine vaccination of smallpox has now been stopped.

Pathogenesis   Studies based on mice and monkeys revealed that usually infection occur by inhalation of virus released from lesions in the oropharynx into oropharyngeal secretions during first week of the rash. After infection of cells in upper or lower respiratory tract, macrophages become infected and enter the lymphatics. Infected white blood cells sometimes enter bloodstream at this stage (viraemia) at about the 4th day. Infected macrophages migrate into epidermis, where they infect nearby cells of basal cell layer. Later there is a migration of polymorphonuclear leucocytes into the lumen of the developing vesicle, which become pustular. The outcome of the infection was either death, which was said to be usually due to 'toxaemia', or recovery with complete elimination of virus with a sequelae of pock marks that occur all over the body (profuse on face).

Clinical manifestations   The incubation period was 10–14 days. The onset was acute, with fever, malaise and headache. Three days after the onset of symptoms, characteristic rashes appeared, intiated on the bucco-pharyngeal mucosa, the face, the forearms and hands, which began as macules that soon became papules, then vesicles and pustular. After 8–9 days of the onset, the pustules became umbilicated and dried up. The characteristic rash were most profuse on the face, more abundant on forearms and lowerlegs, and relatively sparse on the trunk.

Laboratory diagnosis   Diagnosis was by detection of virus particles in negatively stained preparations by electron microscopy (EM), observation of characteristic pocks on the chorioallantoic membrane of the developing chick embryo, detection of nucleic acid by PCR.

HUMAN PARAPOXVIRUSES

Milker's Nodes

It primarily produces localized necrotic lesions on the teats of cows and ewes. In man, it produces localized, painless, non-ulcerating, granulomatous lesions on the hand or face, or both (figure 4.5).

Orf

Orf is an infection in humans caused by virus of contagious pustular dermatitis of sheep and goats. In man, the disease occurs as a single chronic granulomatous lesion with a central ulcer, usually on hand or forearm or occasionally on the face. It heals without scarring.

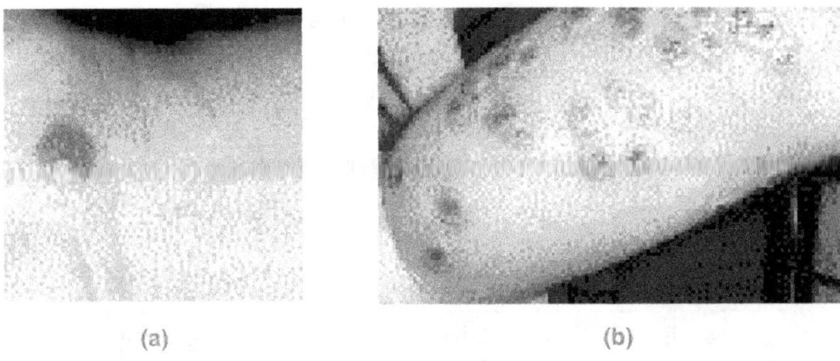

Figure 4.5   Milker's or paravaccinal nodule on the hand is caused by the Orf or parapox virus, family poxvirus.

HUMAN MOLLUSCIPOXVIRUS INFECTION

Molluscum contagiosum is a common, self-limiting viral disease of skin caused by Molluscipoxvirus. It affects only humans; and is characterized by multiple small nodular lesions 2–5 mm in diameter, mostly on the trunk. They become umbilicated and contain white caseous material in which 'molluscum bodies' can be readily demonstrated. These are large (30 μm long) ovoid structures containing many virions. Nodules are dome-shaped, waxy, pearly white or pink in colour and are painless. Lesions can persist for months and usually regress spontaneously. The virus is transmitted by direct contact, perhaps through minor abrasions and sexually in adults. Extensive skin involvement is seen in AIDS patients. The cells in the nodules are greatly hypertrophied and contain large (up to 35 μm) hyaline acidophilic inclusion bodies. These can be seen in cells of the stratum granulosum and the stratum corneum.

HERPESVIRUSES

The name herpes seems to have become attached long ago to this group of viruses because of a rather fanciful idea of the creeping nature (Greek, herpeton, meaning an expert on reptiles) of the lesions caused by some of them. These viruses are excellent examples of 'creepers'.

The family Herpesviridae has been divided into three subfamilies Table 4.1. The virions of this family, have the capacity to establish life long latent infections from which virus may be reactivated. They are frequently reactivated in AIDS and following immunosuppressive therapy for organ transplantation or in cancer.

Table 4.1   Classification of human herpesviruses (HHV).

Subfamily	Virus	Major diseases
Alphaherpesvirinae	HSV-1(herpes simplex virus -1)	Cold sores, fever, blisters, eye and brain infections
	HSV-2 (herpes simplex virus -2)	Genital ulceration
	HHV-3 (Varicella–zoster virus)	Chickenpox, shingles and encephalitis
Betaherpesvirinae	HHV-5 (cytomegalovirus)	Mononucleosis, eye, kidney, brain and congenital infections
	HHV-6	Skin rash, mononucleosis-like illness
	HHV-7	
Gammaherpesvirinae	HHV-4 (Epstein–Barr virus)	Infectious mononucleosis, Burkitt's lymphoma, nasopharyngeal carcinoma

Morphology   Herpesviruses are 120–200 nm in diameter. They comprise four distinct structural elements: envelope, tegument, capsid and core. Envelope is the outermost layer composed of all lipid with numerous small glycoprotein peplomers (figure 4.6). Tegment is the electron-dense material present between the envelope and capsid and it contains several proteins. Inner to the tegument is the icosahedral capsid of diameter 100 nm (figure 4.7). It has a total of 162 tubular capsomeres surrounding a core of DNA. Core consists of a cylindrical protein tegument around which double-stranded DNA is wound. With the exception of Epstein–Barr virus, members of the family *Herpesviridae* can be cultivated in cell cultures and produce giant cells and intranuclear inclusion bodies in infected cells. The genome consists of linear dsDNA with molecular weights varying from 125 to 229 kbp, according to the subfamily.

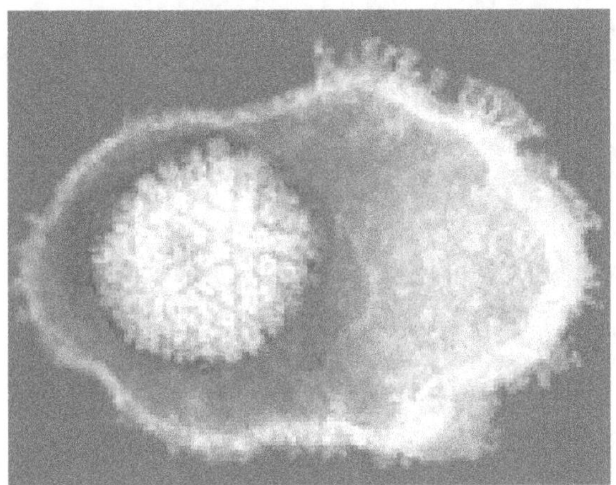

Figure 4.6 HSV.

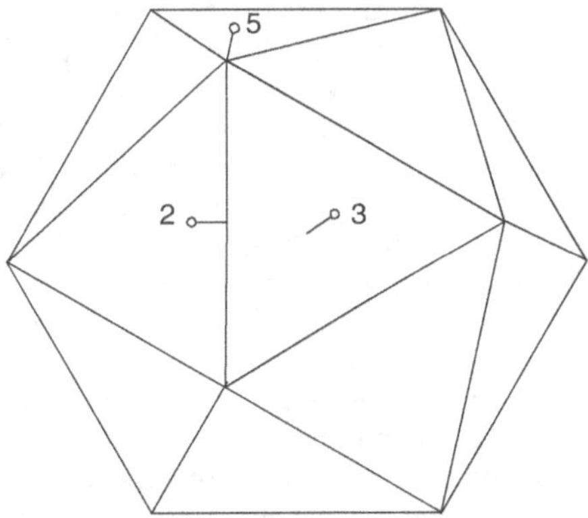

Figure 4.7   The regular icosahedron. This solid has 12 vertices with 5-fold rotational symmetry, the centre of each triangle is on 3-fold axis, and the midpoint of each edge is on a 2-fold symmetry axis. There are 20 identical triangular faces.

HERPES SIMPLEX VIRUS (HSV)

There are two types of herpes simplex virus: type 1 virus (HSV–1) and type 2 virus (HSV–2). HSV–1 infects primarily the mouth,

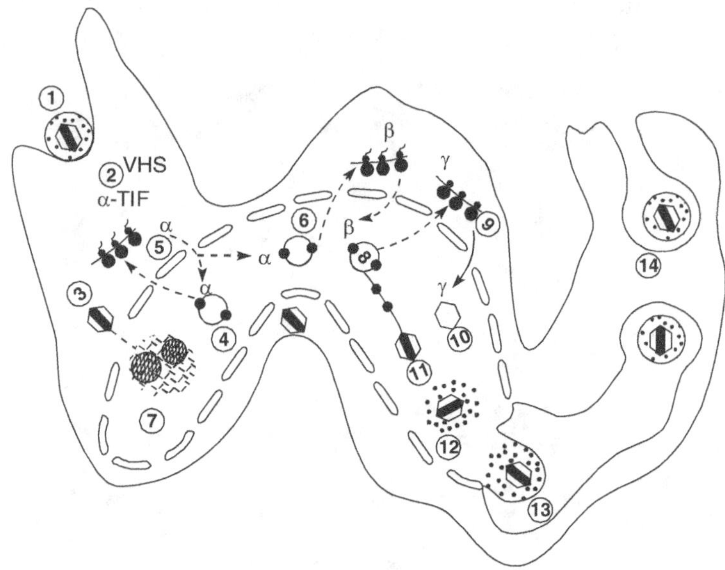

Figure 4.8   Schematic presentation of the replication of Herpes simplex virus. 1. Virus envelope fuses with plasma membrane of cell. 2. Fusion releases 2 proteins, VHS and a-TIF (gene trans inducing factor) 3. Capsid transported to nuclear pore-released DNA circularizes. 4. Transcription of α-genes by cellular enzymes induced by α-TIF. 5. α-mRNA transported into the cytoplasm and translated; proteins enters nucleus. 6. New round of transcription results in β protein synthesis. 7. Chromatin is degraded and displaced toward nuclear membrane 8. Viral DNA replicated by rolling circle mechanism. 9. New round of replication yields γ-proteins. 10. Capsid proteins form empty capsids. 11. Unit length viral DNA is cleared from concatemers and packed into capsids. 12. Capsid with DNA acquire a new protein. 13. Viral glycoprotein and tegument accumulate and form patches in cell membrane. 14. The enveloped virions transported into the extracellular space.

the eye and the central nervous system (regions of the body above the waist), but it is also responsible for a proportion of cases of genital herpes. HSV-2 infects genital and anal regions. The infections caused by herpes simplex viruses can be divided into primary infection, latent infection, reactivation and recrudescence.

Primary Infections

HSV-1 causes acute gingivostomatitis, herpetic whitlow, keratoconjunctivitis, eczema herpeticum, encephalitis and generalized infection.

1. Acute gingivostomatitis   It is the most common primary lesion. It leads to acute and painful ulcers coated with a grayish slough inside the mouth on the buccal mucosa and on the gums. The lesions may also involve the tonsils, pharynx or nose. Normally, the disease is self-limiting and lesions disappear in 2–3 weeks.

2. Herpetic whitlow   It is an occupational hazard of doctors and nurses, who acquire infection by implantation of virus from saliva and respiratory secretions of patients. The lesion is similar to those of staphylococcal whitlow, although the exudate is serous rather than purulent. Vesicles may also be produced on the skin of head and neck.

3. Keratoconjunctivitis   Infection of the eye by HSV-1 causes extremely painful ulceration of cornea and vesiculation of the lids with associated conjunctivitis. In majority of the cases, the primary lesion heals in 2–3 weeks.

4. Eczema herpeticum   It is a superinfection of eczematous skin. It is mainly seen in young children. Crops of vesicles appear mainly on already eczematous areas with extensive ulceration. This results in protein loss, dehydration and viraemia. The latter may lead to disseminated disease with fatal consequences.

5. Encephalitis   Both HSV-1 and HSV-2 can also infect central nervous system leading to herpes encephalitis. The main site of infection is temporal lobe where the disease causes necrosis.

6. *Generalized infections*   Rarely, primary infection with HSV-1 may lead to generalized disseminated infection. Patients develop acute gingivostomatitis, disseminated, vesicular skin lesions, hepatitis and involvement of other organs.

HSV-2 causes genital herpes, aseptic meningitis and neonatal infection. It may rarely cause head and neck infections.

1. *Genital herpes*   HSV-2 causes one of the most prevalent forms of sexually transmitted disease. It leads to the development of painful vesicles on the genitalia or anal regions with fever, malaise and tender, swollen lymph nodes. In the females, lesions may occur on the perineum, vagina, cervix, or vulva. In the male, the lesions may occur on the glans, prepuce or shaft of the penis. HSV-2 proctitis has been reported in homosexual men. Majority (80%) of cases of genital herpes are due to HSV-2 and remaining cases are due to HSV-1. These cases are possibly due to oriogenital sexual practices. HSV-2 may also be involved in oral infections.

2. *Aseptic meningitis*   It may occur as a complication of HSV-2 genital infection.

3. *Neonatal infection*   It is acquired by the neonates usually from their mothers during passage through an infected birth canal, but in some cases it may be acquired in the immediate postnatal period from parents and nurses. Prenatally, very few cases may acquire infections by viraemic transmission across the placenta or by ascending infection from the cervix. This may result in abortions or congenital defects in the child. If a woman has primary genital herpes or neonatal herpes at the time of delivery, then the risk of neonatal herpes is 30–40%. Therefore, caesarean section is indicated in such mothers. Most of the cases of neonatal herpes are due to HSV-2, but those acquired postnatally may be due to HSV-1.

Neonatal HSV infection is invariably symptomatic and frequently lethal. Babies with congenital infection should be identified within 48 h following birth.

Neonatal herpes may be present as

- disseminated disease, with a case fatality rate of 80%, most of the survivors being left with permanent neurologic sequelae, involving CNS, lung, liver, adrenals, skin, eye, etc.

- encephalitis, with high mortality, include seizures, lethargy, irritability, tremors, poor feeding, bulging fontanelle and temperature instability.

- disease localized to mucocutaneous surfaces such as skin, eye and mouth with presentation of discrete vesicles.

Latent infection   During primary infection, the virus travels from the site of infection in the mouth to the trigeminal and probably other cranial and cervical ganglia. Within the sensory ganglia, viral DNA exists as a free circular episome perhaps about 20 copies per infected cell.

Reactivation   Reactivation of the virus is provoked by various stimuli such as common cold, fever, pneumonia, menstruation, exposure to sunlight, stress, etc. Infectious virions migrate along the nerve axon back to the nerve ending, where infection of epithelial cells may result in cluster of vesicles at the mucocutaneous junctions of the lips, nose, or eyes or on areas of skin that have experienced a primary infection. Reactivation recurs sporadically, sometimes often, throughout life.

Laboratory diagnosis   Specimens include vesicle fluid, skin swab, saliva, conjunctival fluid, corneal scrapings, brain biopsy and direct examination of clinical specimens by electron microscopy (EM), fluorescence microscopy (FM) and light microscopy (LM). Herpes virions may be demonstrated in the negatively stained smear of the lesions and tissue preparations. By LM, infected cell may be identified by characteristic changes, which include ballooning of cells, ground-glass nuclear, eosinophilic intranuclear inclusions and multinucleated giant cells. HSV can be isolated on human fibroblast or Vero cells although other mammalian cells also support its growth. Within 1–5 days distinctive foci of swollen, rounded cells appear with some virus strains particularly HSV-2 strains giving rise to fusion of infected cells leading to syncytium formation. Diagnosis can be confirmed within 24 hours by immunofluorescent staining of infected cell culture. Differentiation of HSV-1 and HSV-2 can be made by use of monoclonal antibodies in immunofluorescent staining or by neutralization test with specific antiserum.

Primary infections can be diagnosed serologically by detection of virus-specific IgM or of a rising titre of antibody by complement fixation, neutralization, immunofluorescence, ELISA or RIA. However, serology is not widely used. Polymerase chain reaction may be used for detection of HSV DNA in CSF.

Chemotherapy   HSV infections can be successfully treated with acyclovir (acycloguanosine). It acts by interfering with viral DNA synthesis by inhibiting virus DNA-dependent DNA polymerase. For the treatment of ophthalmic herpes simplex infection, it may be used in the form of ointment. Oral acyclovir 200 mg, 5 times daily for 10 days may be used for the treatment of primary herpes genitalis and orofacial herpes. For the treatment of herpes simplex encephalitis, neonatal herpes and disseminated infection in immunocompromised patients, intravenous acyclovir (10 mg/kg, 3 times daily for 2–3 weeks) may be given. (detailed in Chapter 5)

VARICELLA–ZOSTER VIRUS (VZV)

VZV causes varicella (chicken pox) in children and zoster (shingles) in adults and immunocompromised patients. Varicella follows primary infection in a non–immune individual while zoster is a reactivation of latent virus when immunity has fallen to ineffective level. A child can catch varicella from an elderly patient with zoster, but the latter occurs only if the elderly or immunocompromised person had suffered from varicella in early part of his life.

Varicella

It is one of the common childhood exanthemata. Port of entry of the virus is the respiratory track. Incubation period is about 2 weeks. The earliest manifestation is a flat macular rash which is centripetal (i.e., more pronounced on the trunk than on the limbs) that become raised into papules and progress within a few hours to vesicles. Vesicles are surrounded by a red rim. The lesion then ruptures and the crust may become secondarily infected and pustular before healing.

Pocks are centripetal in distribution, that is they are more profuse on the trunk followed by neck and proximal areas of limbs. Successive crops of vesicles appear over 2–5 days and as a result at any one time, there will be lesions at various stages of development on the same portion of the skin. Some patients of varicella may develop viral pneumonitis, encephalitis, Guillain–Barre syndrome and Reye's syndrome. Varicella tends to be more serious in pregnancy, if the patient has not been

Table 4.2 Salient genetic features of VZV (with 125 kbp genome and 68 ORFs).

Protein	Gene	HSV correlate	Function involved
gpI	68	gE	Most abundant, binds Fc portion of human IgG, molecular weight range of 60-98 kDa
gpII	31	gB	Heterodimer; fusion of envelope, virus entry – 140 kDa protein
gpIII	37	gH	Virus entry – 118 kDa protein
gpIV	67	gI	60 kDa with both N-linked and O-linked oligosaccharide
gpV	14	gC	95–105 kDa protein—Adhesion; low in vaccine strain, virus neutralization with antibody to gpV

(Contd.)

Table 4.2   Contd.

Protein	Gene	HSV correlate	Function involved
gpVI	60	gL	Maturation and transportation of gpIII to cell surface
Nucleocapsid	20,40,41		
Tegument	10	VP16	Transactivator
IE	4	ICP27	Latency, transcriptional activator
IE, tegument	62	ICP-4	Latency; + CMI responses; transcriptional activator
DNA polymerase	28	UL30	

ORF – open reading frame; HSV – herpes simplex virus; gp – glycoprotein; IE – immediate early

infected during childhood. VZV can cross the placenta following viraemia in the pregnant woman and infect the foetus. A syndrome of congenital malformation with hypoplasia of limbs, chorioretinitis and scarring of skin associated with maternal varicella may develop in the first trimester.

Zoster

Zoster or shingles is mainly a disease of adults, and is an endogenous reactivation of virus which has remained latent in one or more sensory ganglia following primary varicella infection many years earlier. 'Zoster' is derived from the Latin word for a "belt" or "girdle" and refers to the characteristic distribution of the rash when a thoracic dermatome is involved. Virus travels down the sensory nerves to produce painful vesicles in the area of skin (dermatome) enervated from the affected ganglion. Thoracic nerves supplying the chest wall are most often affected when the ophthalmic nerve of trigeminal ganglion is affected. The rash is distributed on the scalp and forehead. In about half the patients, the eye is affected leading to corneal ulceration and stromal keratitis.

The accompanying pain is often very severe for up to a few weeks, but herpetic neuralgia, which occurs in half of all patients over 50 years of age, may persist for months which may require surgical ablation of the ganglion zoster of the seventh cranial ganglion may lead to Bells palsy which is characterized by facial nerve palsy with a rash on the tympanic membrane and the external auditory canal. Zoster may also cause encephalitis.

Congenital varicella syndrome   Characterized based on developmental abnormalities described in case reports of infants born to mothers who had varicella in early pregnancy. Defects include hypoplasia of lower limb, eye abnormalities (Chorioretinitis, microphthalmia, Horner syndrome, cataract and nystagmus), skin scars (cicatricial), cortical atrophy and for mental retardation and early death.

Laboratory Diagnosis

1. Direct examination of vesicle fluid by electron microscopy reveals herpes particles.

2. Stained smears from the base of the lesion or sections from biopsy tissue show multinucleated giant cells containing acidophilic intranuclear inclusion bodies.

3. Rapid diagnosis is possible by using monoclonal fluorescent antibody technique.

4. VZV antigens can be detected in vesicle fluid by ELISA .

5. DNA can be extracted from virions in vesicle fluid, amplified by PCR and detected by nucleic acid hybridization.

6. The virus can be isolated from vesicle fluid in human embryonic lung fibroblast culture. Characteristic cytopatheic effect consists of foci of swollen refractile cells in a spindle shaped configuration. It develops slowly over a period of 2 or more weeks. However, VZV antigen can be demonstrated in nuclear inclusion by

immunofluorescence with monoclonal antibody before the end of the first week.

7.   Recent infection can be diagnosed by ELISA test for varicella–zoster specific IgM antibody in the patient's serum.

Treatment

Acyclovir and vidarabine given intravenously (10 mg/kg every 8 hrs) are effective in the treatment of severe varicella–zoster (e.g. in the immunocompromised patients) infections.

EPSTEIN–BARR VIRUS (EBV)

EBV has been named after the virologists (Epstein and Barr) who first observed it under electron microscope in cultures of lymphoblasts from Burkitt's lymphoma named after the British surgeon Denis P. Burkitt. EBV replicates in epithelial cells of nasopharynx and salivary glands, especially the parotid, lysing them and releasing infectious virions into saliva. B lymphocytes appear to become infected when they infiltrate infected nasopharyngeal mucosa. Inside the B cells, EBV normally fails to replicate but establishes lifelong latency. In these cells, EBV genome persists as multiple full-length copies in the form of circular episomes. EBV has oncogenic properties and a proportion of infected B lymphocytes undergo transformation and transform cells in vitro too. Activated B lymphocytes secrete immunoglobulins especially IgM.

The genome of EBV consists of a linear 172 kbp, dsDNA with tandemly reiterated 0.5 kbp terminal direct repeats (TR), and tandemly reiterated 3 kbp internal direct repeats (IR) that divide the genome into unique long (UL) and short (US) regions.

EBV possesses envelope glycoprotein gp 350/220 which mediates attachment of the virus to CD21 receptors present on the susceptible cells (B-cells). Most shedding of virus takes place in the oral cavity, therefore transmission of virus requires salivary contact, either through kissing or contaminated eating utensils. Most infections are symptomless, especially when acquired during childhood however in adolescents it may cause infectious mononucleosis or glandular fever, infections in immunocompromised hosts and when EBV exerts its oncogenic potential it may cause Burkitt's lymphoma and nasopharyngeal carcinoma.

Virus expression in latent infection is characterized by three distinct processes i) viral persistence ii) restricted virus expression that alters cell growth and proliferation, and iii) retained potential for reactivation to lytic replication.

Most infected cells carry many copies of EBV genome, which exist as closed circular dsDNA plasmids (episomes). However, it cannot be ruled out that some small fragment of the EBV may be integrated into the host cell genome to cause the "stable" transformation of lymphocytes that leads to continuous growing cultures.

The 172 kbp EBV genome encodes ~100genes, 10 of which are expressed during latency and transformation including 6 EBV nuclear binding proteins (EBNAs 1, 2, 3A, 3B, 3C, LP) of 150–200 kDa with diverse function, two latent membrane proteins (LMP-1 and -2), and two small untranslated, non-polyadenylated RNAs (EBERs 1 and 2). EBNA-1 expression appears to be associated with transformation of B-cells

Table 4.3   EBV growth transformation associated antigens.

Protein	Subcellular location	DNA binding	Size	Function involved
EBNA-1	Nucleus	Yes	70–92 kDa	Replication and maintenance of episome
EBNA-2	Nucleus	Yes	86–89 kDa	Initiation of transformation, expression of other EBV latent genes, transactivator of cellular and EBV genes

(Contd.)

Table 4.3   Contd.

Protein	Subcellular location	DNA binding	Size	Function involved
EBNA-3 (3A, 3B, 3C)	Nucleus	Yes	145–165 kDa	Promotes virus transformation competent in B-cells growth
EBNA-LP	Nucleus	Yes		May play a role in RNA processing or association with some nuclear regulatory protein
LMP-1	Cytoplasmic membrane	No	63 kDa	Factor needed for tumorigenic conversion of keratinocytes
LMP-2 (2A, 2B)	Cytoplasmic membrane	No		Association with tyrosine kinase may influence EBV's effect on cell growth

EBNA – EBV nuclear antigen; LP- leader protein; LMP – latent membrane protein

(i.e., Burkitt's Lymphoma) and nasopharyngeal carcinoma (NPC) The antigens associated with EBV growth transformation are listed in Table 4.3.

The presence of clastogenic (chromosome breaking) substances during EBV infection has been reported, and presence of EBV DNA has been observed in epithelial cells of nasopharyngeal carcinoma.

Infectious Mononucleosis IM (Glandular Fever)

It is a primary EBV infection seen mainly in the 15–25 years age group. After an incubation period of 4–7 weeks, patients encounter sore throat due to exudative tonsillitis, generalized lymphadenopathy, fever, malaise, headache, sweating, fatigue and gastrointestinal discomfort. These features in association with lymphocytosis caused by presence of atypical lymphocytes constitute the illness referred to as IM. In some cases, spleen and liver are often enlarged. A faint transient morbilliform rash may be seen. A maculopapular rash may appear, especially following treatment with ampicillin due to immune complexes with antibody to ampicillin. The disease usually lasts for 2–3 weeks. However, some patients develop complications such as Guillain–Barre syndrome, Bell's palsy, meningoencephalitis, transverse myelitis, haemolytic anaemia, thrombocytopenia, carditis, nephritis, pneumonia and splenic rupture.

Infections in the immunocompromised include X-linked lymphoproliferative syndrome in male members of certain families with an X-linked recessive immunodeficiency with a reduced ability to synthesize interferon. About half of the boys die from sepsis or haemorrhages within a month, whereas the remaining die from B cell lymphoproliferation associated with hypogammaglobulinaemia and later lymphoma.

EBV may cause progressive lymphoproliferative disease in transplant recipients, immunodeficient children and AIDS patients.

EBV–Associated Malignancies

1. Burkitt's lymphoma   This lymphoma is a malignant B-cell lymphoma seen in regions of equatorial Africa and New Guinea where malaria is hyperendemic. EBV infection is acquired early in life. Malaria being immunosuppressive, may act as a cofactor by interfering with the immune responses that normally prevent reactivation of EBV from B cells thus enabling its full oncogenic potential to be expressed. The EBV genomic DNA is present in multiple copies in the form of circular episomes, in each cell of most African Burkitt's lymphomas, but the cells do not produce virus unless induced to do so following cultivation in vitro.

2. *Nasopharyngeal carcinoma*   It is also associated with EBV and virus DNA is regularly present in the malignant epithelial cells of the tumour. It shows a striking geographical and probably racial distribution, for example, the disease is common in people from southern China, Eskimos and Greenlanders. Affected cells express a specific subgroup of EBV latent proteins, including EBNA-1, LMP-1 and LMP-2. LMP-1 is a likely prerequisite to this neoplastic transformation.

3. *B cell lymphoma*   Immunodeficient patients, e.g. recipients of transplants and HIV infected patients may develop EBV-associated B-cell lymphoma.

4. *Oral hairy leukoplakia*   EBV is believed to cause oral hairy leukoplakia in HIV-infected patients. Capsids of EBV can be seen tissue in epithelial cells of affected tissue located along the side of the tongue in AIDS patients.

5. *Lymphocytic interstitial pneumonitis*   EBV has also been associated with lymphocytic interstitial pneumonitis in HIV-infected children.

Laboratory diagnosis

1.   *Differential white blood cell count*   By the second week of the illness patient develops leucocytosis (10,000–20,000/cu.mm or more). Lymphocytes and monocytes account for 60–80%. Of these, 20% or more are "atypical lymphocytes." The latter are large pleomorphic blasts with deeply basophilic vacuolated cytoplasm and lobulated nuclei. They persist for 2 weeks to several months.

2.   *Paul–Bunnel heterophile antibodies*   Infectious mononucleosis is accompanied by production of heterophile agglutinins. These are heterophile IgM antibodies elicited by EBV infection. These antibodies appear in 85–90% of patient's sera during the acute phase of illness, reaching peak levels 2 weeks after the onset. Their titre decreases rapidly after fourth week and are not detectable after 3 months. Heterophile antibodies may be readily detected by a rapid slide agglutination test or Paul–Bunnell

test. Agglutination of sheep or horse red cells by patient serum, adsorbed with guinea pig kidney cells to remove Forssman antibody, is the basis of this test.

3. *EBV-specific antibodies*   More reliable indicator of EBV infection is the demonstration of IgM antibody to the EBV viral capsid antigen by ELISA or indirect immunofluorescence. This antibody becomes detectable within 4 weeks and declines rapidly over the next 3 months or so.

4. *Virus isolation*   Saliva or throat washing and peripheral blood leucocytes can be inoculated onto umbilical cord lymphocytes. If specimen contains EBV, it leads to immortalization of the cells to produce a lymphoblastoid cell line. EBV can also be detected by PCR and DNA hybridization. In biopsy of affected tumours, viral DNA can be demonstrated by staining with specific antibodies.

Treatment   Supportive care with Acetaminophen or non-steroidal anti-inflammatory agents. Antiviral drug Acyclovir inhibits permissive EBV infection through inhibition of EBV DNA polymerase but has no effect on latent infection.

CYTOMEGALOVIRUS (CMV)

The name means "large cell virus" and derives from the swollen cells containing large intranuclear inclusions that characterize these infections. CMV is transmissible to the foetus via the placenta, and is an important cause of neonatal morbidity and mortality. Normal infants can acquire infection from colostrums or breast milk. In vivo, CMV replicates in epithelial cells in salivary glands, kidneys and respiratory epithelium. In vitro, they can be isolated on human fibroblast cells. An individual infected with CMV carries the virus for life and may shed it intermittently in saliva, urine, semen, cervical secretions and breast milk. The virus is found in 0.3–2.4% of the population that have been sampled throughout the world. Infection is transmitted by close contact and by blood transfusion. It may be acquired at any time, i.e., prenatal, perinatal and postnatal.

CMV proteins   Presentation of protein mapping data follows "The international CMV workshop in 1993 agreed on a nomenclature for the description of CMV proteins". The system designates 'p' for protein, 'gp' for glycoprotein, 'pp' for phosphoprotein, followed by the genetic locus and any preferred trivial name in brackets. For instance, gpUL55 (gB) is the glycoprotein encoded by the 55th ORF in the unique long region, known colloquially as glycoprotein B (gB).

Prenatal infection   CMV is the most common agent to cause intrauterine infection and prenatal damage to foetus leading to congenital abnormalities. Approximately 1% of all babies become infected in utero. Maternal viraemia following primary CMV infection or a reactivation during pregnancy may result in foetal infection. Majority (95%) of these is without obvious symptoms at the time of birth and 5% symptomatic infants have cytomegalic inclusion body disease. These infants show signs of growth retardation, hepatosplenomegaly, jaundice, thrombocytopenia, microcephaly, encephalitis and chorioretinitis. Of the remaining 95%, about 15% go on to develop deafness and mental retardation.

Perinatal infection   This is acquired from infected maternal genital secretions or from breast feeding.

Postnatal infection   This may be acquired by kissing (from saliva), sexual intercourse or artificial insemination (from semen),blood transfusion and organ transplantation. Infection acquired after birth are generally subclinical. However, it may cause hepatitis in young children. In adults and older children, it may cause a syndrome resembling EBV infectious mononucleosis, but with a negative Paul–Bunnell test and no pharyngitis or lymphadenopathy. CMV may cause widely disseminated infection in immunocompromised individuals such as graft recipients and AIDS patients leading to interstitial pneumonia, chorioretinitis, hepatitis, arthritis, gastrointestinal infection, encephalitis, Guillain–Barre syndrome and transverse myelitis.

Capsid proteins		
pUL86	Major capsid protein	Cross react with VP5 of HSV and share epitopes with *gag* precursor p55 from HIV
pUL46	Minor capsid protein	Homology with VP19c of HSV
pUL80a (assemblin)	Assembly protein	Analogy with VP22a of HSV, autocleavage occur at three sites for polyprotein processing
Tegument proteins		
ppUL32	Basic phosphoprotein	Elicits immunodominant serologic response in humans
ppUL83	Lower matrix protein pp65	Component of dense bodies and target for class I restricted cytotoxic T lymphocytes
ppUL82	Upper matrix protein pp71	Potent transactivator
ppUL56		DNA binding protein, Immunogenic
ppUL99 (pp28)		Immunogenic, constituent of tegument
Envelope proteins		
gpUL55	Glycoprotein B, gp complex I	Has homology to gB of HSV; has two continuous neutralization epitopes antigenic determinants— AD1 and AD2
gpUL75	Glycoprotein H; gp complex III	Involved in binding of the virus to the cell surface or internalization
gpUL115	gpL	Act as chaperone to facilitate display of gH
gpUL100	Integral membrane protein; gp complex II	Have multiple domains; has homologs in HSV, VZV and EBV

Laboratory diagnosis   CMV can be isolated from urine, saliva, breast milk, semen, cervical secretions and blood leucocytes. These are inoculated on cultured human fibroblasts. The virus replicates very slowly, therefore, characteristic CPE of foci of swollen refractile cells with cytoplasmic granules may take 2–3 weeks to appear. When stained, these cells are multinucleated giant cells containing acidophilic inclusions in the nuclei and cytoplasm. For precise identification, the cell line may be stained by immunomonoclonal or immunoperoxidase technique using monoclonal antibody. Antigen detection can be done by DEAFF test (for detection of early antigen fluorescent foci. CMV DNA, in the specimen, can be amplified by PCR. CMV-specific IgM can be detected in the patient serum by ELISA.

Treatment   For the treatment of severe CMV infections such as pneumonia, chorioretinitis and colitis in AIDS patient or in other immunocompromised patients, ganciclovir is the drug of choice.

Human Herpes virus 6(HHV-6)   The agent was first isolated in 1986 from blood leucocytes of six patients suffering from lymphoma or acute lymphocytic leukaemia. Infected cells show ballooning with nuclear and/or cytoplasmic inclusions in human B lymphocytes. Macrophages comprise an important reservoir of HHV-6. Saliva acts as the main route of transmission of this virus. It can be cultured on transformed B-cells, NK cells, giant cells, fibroblasts and epithelial cells. It causes febrile illness of children, *Exanthema subitum* or *Roseola infantum*, which is a mild facial rash occurring commonly between 6 months and 3 years of age with sudden onset of fever. Secondly, Mononucleosis with cervical lymphadenopathy in few adults during primary infection.

Laboratory diagnosis include detection of antigen by immunofluorescence (IF) using monoclonal antibodies (MAbs) and genome detection by PCR. Serology is done by using (Enzyme Linked Immunosorbent Assay (ELISA). HHV-6 can be isolated from peripheral blood mononuclear cells in early febrile stage of the illness by co-cultivation with cord blood lymphocytes.

Human Herpes virus 7 (HHV-7)   HHV-7 was isolated from CD4+ T cells taken from a healthy person in 1990. Unlike HHV-6, HHV-7 may cause an illness resembling infectious mononucleosis, but the results of primary infections are not as clear as those caused by HHV-6.

Human Herpes virus 8 (HHV-8)   HHV-8 was identified in 1994 and is suggestive of a causal relation in Kaposi's sarcoma that is common in HIV patients. Laboratory diagnosis of HHV-8 includes its DNA amplification by PCR. Cutaneous manifestations of human virus infections are shown in Table 4.4.

Table 4.4   Cutaneous manifestations of viral diseases.

Virus family	Prototype virus	Primary cutaneous lesion
DNA viruses		
Poxviridae	Molluscum contagiosum virus	Papules
	Variola virus	Papules, vesicles, pustules
	Orf virus	Papules, vesicles
Herpesviridae	Herpes simplex virus, Varicella virus	Vesicles
	Epstein–Barr virus	Oral—plagues (AIDS patients), macules (intramuscular)
	Cytomegalovirus	Ulcers (AIDS patients), macules, papules (in TORCH syndrome cases)
	Human herpes virus 6	Macules (Roseola)
	Human herpes virus 8	Nodules
Papovaviridae	Papillomavirus	Papules, nodules
Parvoviridae	Parvovirus B19	Macules, papules
RNA viruses		
Picornaviridae	Coxsackie virus (Herpangina)	Vesicles
Paramyxoviridae	Measles virus	Macules, papules
Togaviridae	Rubella virus	Macules
Retroviridae	HIV	Papules

ADENOVIRIDAE

Members of the family *Adenoviridae* are non-enveloped icosahedral viruses containing linear double-stranded DNA that replicate in the nucleus of the infected cell. The family comprises two distinct genera: Mastadenovirus and Aviadenovirus, i.e., mammalian and avian adenoviruses respectively. There are 47 serotypes of human adenoviruses, which have been assigned to 6 (A–F) subgenera. They can be found in healthy persons the tonsils or adenoid glands (hence *adeno* from the Latin for "glands").

Morphology

Adenoviruses are icosahedral virions containing dsDNA. They measure 80 nm in diameter. Each capsid is composed of 252 capsomeres: 240 hexons make up the 20 triangular faces of icosahedron and 12 pentons form the vertices. From each penton projects an apical fibre, 9–31 nm in length and of 62 KDa that serves to bind specifically to receptor sites on the host cell (figure 4.9).

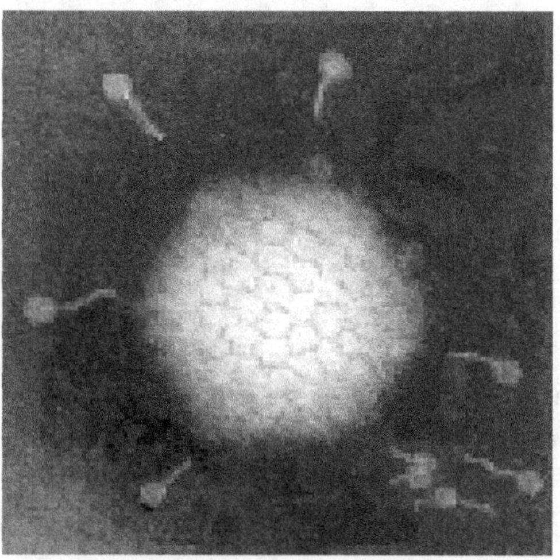

Figure 4.9   Adenovirus.

Properties

Resistance   Adenoviruses remain viable for about a week at 37°C. They are readily inactivated at 50°C. They resist ether and bile salts.

Cultivation   Human adenoviruses can be grown on monolayers of HeLa, HEp-2, KB and human embryo kidney cells. Cytopatheic effects may take 1–4 weeks and consists of cell rounding and aggregation in grape-like clusters. Infected cells swell and become ballooned and show characteristic basophilic intranuclear inclusion in stained preparation.

A 100 kDa polypeptide provides the base necessary for the assembly of hexons. Viral group specific antigen is located on the inner surface of the hexon capsomer, type specific antigen is found on other region of hexon and on the fibre. Fibre determinant (γ) is a type specific hemagglutinin. Main internal core protein include ppV(48 kDa) and ppVII (18 kDa). DNA of adenovirus is linear, dsDNA with a molecular weight range of 20×10^3 to 24×10^3 kDa for different serotypes. DNA shows terminal inverted repeats of more than 100 bp, enabling the DNA, when single-stranded, to form panhandle-shaped molecules (figure 4.10).

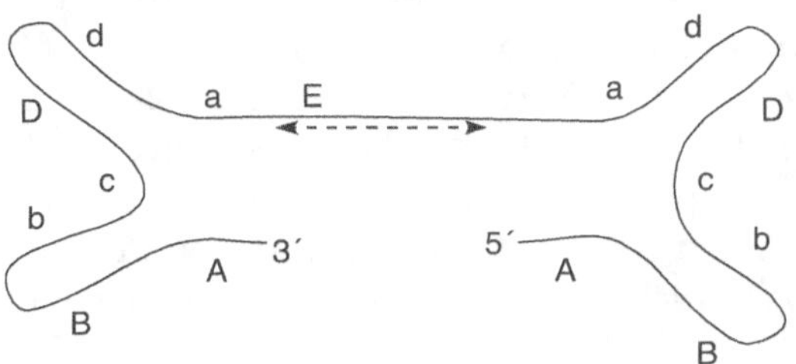

Figure 4.10   Panhandle-shaped molecules of DNA.

ABCDbaE,F: 'AB' are nucleotides 1–41 that are complementary to 'ab' (nt 125–85), C and D are two shorter self complementary hairpin sequence (nt 42–62 and 64-84 respectively), E, F represent the continuous DNA sequence.

Adenovirus Replication

Adenovirus fibre attaches to a specific receptor (integrins) on the cell surface. Virion is then endocytosed as a receptosome at low pH (5.0). Virus loses its penton capsomer in the cytoplasm transferred to the nucleus via microtubules. Upon entering the nucleus viral DNA becomes coated with protecting histones into a complex. It then undergoes early transcription with the help of host RNA polymerase II, prior to any DNA replication.

DNA replication starts at the 5´ and 3´ ends on both positive (right strand) and negative (left strand) sense strands. A separated copy of the DNA strand serves as a template for the completion of dsDNA molecule. Viral DNA encodes a total of 8 ORFs (open reading frames), 6 ORFs are active early after infection (E1A, E1B, E2A, E3, E4, L1), and three become activated at intermediate (PIX, IVa2) and late time of infection (late ORF).

The whole adenovirus replication cycle is 24–30 h for different serotypes. Early mDNA appear during the first 8–10 h. The positive strand gives rise to three important products, one of which induces DNA synthesis, second one acts as transactivator for other viral promoters, third one interacts with cellular oncogenes (*c-ras*) to produce fully transformed phenotypes. Adenoviruses give rise to tumour (T) associated antigen (17–58 kDa). T antigens are products of early genes E1A, E1B, E2A codes for a 22 kDa DNA binding phosphoprotein, E2B from negative strand codes for a 72 kDa ssDNA binding protein for viral DNA synthesis. Late protein IVa2 is a scaffolding protein and transcription factor augmenting major late promoter activity. Most viral proteins are rapidly transported to the cell nucleus for virus assembly after synthesis.

Pathogenesis

Adenoviruses may infect via the conjunctiva or the nasal mucosa. Faecal-oral spread, particularly among children can also occur. They multiply initially in the conjunctiva, pharynx or small intestine, spread to preauricula, cervical and mesenteric lymph nodes. Most of the enteric and some of the respiratory infections are subclinical. In children, asymptomatic infection of tonsils and adenoids, leading to respiratory carriage, and Peyer's patches, leading to gut carriage, may persist for weeks or months.

Most of the adenovirus infections are caused by adenoviruses serotypes 1–8. Serotypes 40 and 41 may cause infantile gastroenteritis 8, 19 and 37 lead to eye infections and, serotypes 19 and 37 may also cause genital infections. Five serotypes (43–47, subgenus D) have been recovered from the faeces of AIDS patients. However, their role in these patients is not known. Adenovirus infections lead to lasting immunity to reinfection with the same serotype.

Clinical Syndromes

Incubation period is 5–8 days after which it may lead to:

Respiratory infections

1. Tonsillitis particularly in young children 3 years of age. Serotype 1–7 are usually responsible for this infection.

2. Pneumonia in infants and young children may be caused by any of the common serotype particularly 1, 2, 3 and 7. In military recruits, pneumonia may be caused by serotypes 4 and 7.

3. *Acute respiratory disease*   This syndrome is characterized by fever, chill, pharyngitis, cervical lymphadenitis, non-productive cough and malaise. This occurs usually in outbreaks in military recruits when they assemble in camps. It is generally caused by adenovirus serotypes 3 and 7.

Ocular infections

1. *Pharyngoconjunctival fever*   This syndrome of febrile pharyngitis and conjunctivitis occur in outbreaks in

children. It is usually caused by adenovirus serotypes 3, 4, 7 and 14.

2. *Epidemic keratoconjunctivitis*   It is a severe and highly contagious infection involving all groups. It often occurs in epidemic form and is characterized by follicular conjunctivitis and progresses to involve the cornea. It is usually caused by adenovirus serotypes 8, 19 and 37.

Genitourinary infections

1. *Cervicitis and urethritis*   These venereal infections are caused by serotype 37.

2. *Acute haemorrhagic cystitis*   It is mainly seen in infants and young children and is caused by serotypes 11 and 21.

Enteric infections   Serotypes 40 and 41 may cause infantile gastroenteritis in up to 10% of the cases. Some human adenoviruses produce carcinomas if injected into newborn hamsters, mice or rats. Infection of cultured hamster or rat cells, in vitro, also results in transformation. However, there is no evidence that this can occur in man too.

Laboratory Diagnosis

Specimens for the laboratory diagnosis of adenovirus infections include stool, throat swab, nasopharyngeal aspirate, bronchial lavage, conjunctival swab, corneal scraping, urine, genital secretions and biopsy and autopsy materials.

- Virus particles may be seen directly in stool extracts by electron microscopy.

- Viral antigens in the cells from respiratory tract, eye, urine, biopsy or autopsy material and infected cell cultures A549, HEP-2, KB and MRC-5 may be detected by immunofluorescence using polyclonal or monoclonal antibodies.

- Viral antigens in faeces and nasopharyngeal secretions may be detected by ELISA using monoclonal or polyclonal antibodies directed against the group-specific hexonantigen.

- Enteric adenoviruses may be detected by latex agglutination method using latex particles coated with specific antibody to each virus.

- Viral DNA in the faeces may be detected by nucleic acid hybridization and PCR.

- Virus from the clinical specimens may be isolated on HeLa, HEp-2, KB and human embryo kidney cells.

Treatment

At present there is no specific antiviral treatment for adenoviral infections. Ribavirin is active against adenoviruses in vitro. Ribavirn is available for i.v. use. Cidofovir is active against adenoviruses in vitro, but clinical studies are lacking.

PARVOVIRIDAE

Family Parvoviridae comprises very small (20–25 nm) non-enveloped, icosahedral virus containing a linear single-stranded DNA of $\sim$1–5 $\times$ 10^3 kDa. The parvoviruses can replicate only in the presence of another virus or of active DNA synthesis in rapidly dividing host cells. The reason lies in their minute size, for these are the smallest of all viruses (Latin, parvus meaning small). There are three genera in this family: Parvovirus, Erythrovirus and Dependovirus (Table 4.5). Members of the genus Dependovirus are defective and require co-infection with a helper virus for their own replication. They are usually found in association with an adenovirus and are known as "adeno-associated viruses" (AAV). They have not yet been associated with any human disease. Genus Parvovirus is capable of replication without the aid of a helper virus. They cause disease in animals but some may cause diarrhoea in humans. Parvovirus B 19 has now been allocated to a separate genus, Erythrovirus.

Table 4.5 Parvoviruses infecting humans.

Genus	Viruses	Diseases
Erythrovirus	B19	Erythema infectiosum Foetal infections Aplastic crises
Parvovirus	RAV-1	Possibly implicated in rheumatoid arthritis
Dependovirus	Adeno-associated viruses AAV) types 1–5	AAV-2 possibly implicated in foetal infections

Morphology

These are non-enveloped viruses with icosahedral symmetry, about 22 nm in diameter (figure 4.11).

Genome

The parvoviridae are the only DNA viruses with single-stranded nucleic acid. A 'hairpin' loop at the end of each genome initiates replication of the complementary strand. The ssDNA is about 5 kb long.

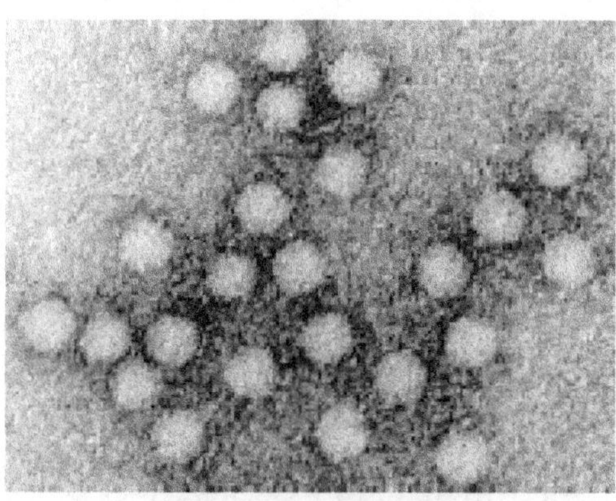

Figure 4.11 Parvovirus.

Replication

The virus attaches to a globoside or P-antigen receptor, is endocytosed, and migrates to the nucleus where replication of virus DNA and transcription of mRNA takes place. Nine viral mRNA species and up to seven viral proteins are detectable. The dependoviruses are used to be called "adeno-associated viruses" since they were seen by EM in association with enteric adenoviruses. The genome becomes integrated into host-cell DNA and is active only when the cell becomes infected with a helper virus that can provide the enzymes essential for replication.

Erythrovirus (Parvovirus B19)

Parvovirus B19 is highly contagious. It is transmitted by respiratory secretions and close contact. Other routes of transmission are transplacental and via blood transfusion. It causes erythema infectiosum, chronic anaemia, arthropathy, transient aplastic crisis and erythroblastosis foetalis.

The clinical syndromes of the various diseases are discussed in detail below.

1. *Erythema infectiosum*   It is a common childhood febrile illness with rash. It is characterized by an erythematous skin eruption of the cheeks ("slapped cheek") and a reticulated lace-like eruption on the trunk and extremities. The rash usually resolves in 1–2 week. The rash may be preceded by fever, headache, malaise and myalgia.

2. *Chronic anaemia in immunodeficient patient*   Patients with acute leukaemia on chemotherapy, AIDS patient, children with congenital immunodeficiency syndrome and bone marrow transplant recipients if infected with parvovirus B19, may lead to chronic anaemia.

3. *Arthropathy*   Susceptible adults may develop arthritis or arthralgia.

4. *Transient aplastic crisis*   Parvovirus B19 has an affinity for immature red blood cell precursors. It can lead to transient aplastic crisis in patients with a variety of

conditions that require increased red cell production, including sickle-cell anaemia, thalassemia, hereditary spherocytosis and acute blood loss.

5. *Erythroblastosis foetalis* If a pregnant woman develops primary parvovirus B19 infection, fatal erythroblastosis foetalis with hydrops may result.

Laboratory diagnosis

1. *Virus isolation and detection* Parvovirus B19 may be cultured in cells from human bone marrow or foetal liver in the presence of erythropoietin and interleukin-3. By electron microscopy, it can be detected in the patient serum.

2. *Nucleic acid detection* Viral nucleic acid can be detected by nucleic acid hybridization or by polymerase chain reaction.

3. *Antigen detection* Antigen can be detected in serum and other specimens by ELISA or RIA. In the infected cells, the antigen may be detected by indirect immunofluorescence or by immunoperoxidase staining with monoclonal or polyclonal antibodies.

4. *Antibody detection* Most successful diagnostic assay is the detection of IgM antibodies or a significant rise in IgG antibodies by ELISA or RIA.

PAPOVAVIRIDAE

The name of this family is yet another example of the acronyms so beloved by virologists:

Pa = papillomaviruses

Po = polyomaviruses

Va = vacuolating agents

The-*oma* components of these names immediately alert you to the fact that these agents have something to do with neoplasia (Greek, oma, meaning tumour) and, indeed, the papillomaviruses have now displaced herpes simplex as the centre of attention in relation to cancers of the genital tract.

The family *Papovaviridae* constitutes a group of viruses containing a small icosahedral capsid composed of 72 capsomeres and super coiled dsDNA genome. It can be divided into two genera. Papillomavirus and Polyomavirus. They induce both lytic infections and either benign or malignant tumours.

Papillomaviruses

Papillomaviruses are 55 nm in diameter (figure 4.12) and have an icosahedral capsid composed of 72 capsomeres (pentamers). The genome is a supercoiled dsDNA molecule of 8 kbp. The genome comprises of eight ORFs, all located on the same stand. These have been designated as either "early" (E) or "late" (L). The early ORFs (E1, E2, E4, E5, E6, E7) code for non-structural, regulatory proteins. The E3 ORF does not code for a protein, and the E8 ORF has been identified only in bovine papilloma viruses. The late ORFs (L1 and L2) code for capsid proteins. A 1kb-long upstream regulatory region (URR) lies between the E and L ORFs, which include the origin of replication (ori). They are host species-specific and infect the squamous epithelia and mucous membranes of higher vertebrates, including man. There are over 62 types of human papillomaviruses (HPV). These are distinguished on the basis of the homology between their genomes. Viral proteins E5, E6, and E7 are involved in malignant transformation of cells; and E1 and E2 participate in DNA replication.

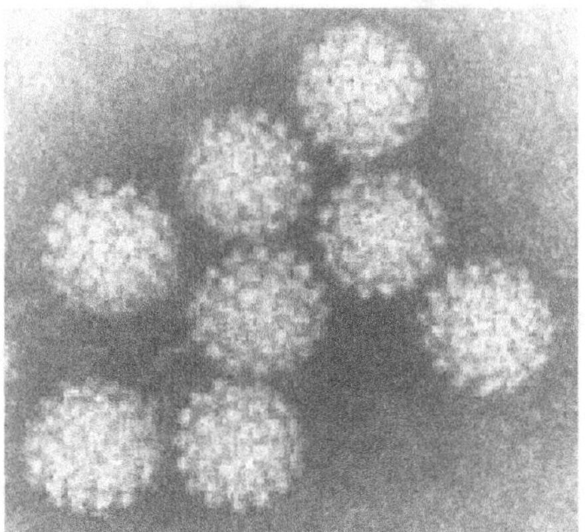

Figure 4.12   Papillomavirus.

Clinical features   Papillomaviruses are classified on the basis of their degree of DNA homology. Nearly 80 types of HPV have now been identified; they cause disease only in skin and mucous membranes, where they give rise to warty lesions (Table 4.6).

Pathogenesis   HPV are not only host species-specific but also display a predilection for the skin and others for the mucous membrane. They cause cutaneous warts, genital warts, recurrent respiratory papillomatosis, oral papillomatosis and cancer. The infection is transmitted by indirect or direct contact including sexual contact.

1. *Cutaneous warts*   These are generally caused by HPV types 2, 4 and 7. Virus penetrates the skin through an abrasion and infects the basal cell layer leading to hypertrophy of all the layers of the dermis and hyperkeratosis of the horny layer. Incubation period is 2 years. Cutaneous warts are raised papillomas with a rough surface and are commonly found in children. They disappear within 2 years of the onset. Recurrence is believed to be due to persistence of the virus in the skin surrounding the original wart.

Table 4.6   Types of papillomavirus lesions.

Type of lesion	Site	HPV types
Predominantly benign lesions		
Common warts	Skin, various sites	2, 4
Plantar and palmar warts	Hands, feet	1, 2, 4
Butcher's warts	Hands	7
Flat warts	Skin, various sites	3
Genital warts (Condyloma acuminate)	Cervix, vulva, penis	6, 11
Juvenile laryngeal papilloma	Larynx	6,11

(Contd.)

Table 4.6   Contd.

Type of lesion	Site	HPV types
Malignant or potentially malignant lesions		
Flat warts	Skin	10
Bowenoid papulosis	Vulva, penis	16
Malignant intraepithelial neoplasia	Cervix, penis	11, 16, 18, 39–45, 51–56
Carcinoma	Cervix, penis	16, 18
Papilloma/carcinoma	Larynx	16
Epidermodysplasia verruciformis	Skin, various sites	5, 8, 10, 14, 15, 17, 19–29

2. *Genital warts*   These lesions are also known as condyloma acuminata and are common in sexually active adults. In women they are found on the vulva, within vagina and on the cervix. In men, the lesions generally appear on the shaft of the penis, perianal skin and anal canal. Genital warts are generally caused by HPV types 6 and 11.

3. *Recurrent respiratory papillomatosis*   HPV types 6 and 11 may lead to benign squamous papillomas on the mucosa of respiratory tract, particularly on the larynx. Children below 5 years of age acquire the disease by passage through an infected birth canal. Adults after the age of 15 may acquire the disease by orogenital contact with an infected sexual partner. Malignant conversion of laryngeal papillomas has been reported.

4. *Oral papillomatosis*   Infection is usually acquired by orogenital contact with infected sexual partner. It is caused by HPV types 6, 7, 11, 13, 16 and 32. Multiple papillomatous lesions may develop on the buccal mucosa, a condition known as oral florid papillomatosis. Oral papillomas sometimes progress to malignancy. HPV-16 DNA has been reported in a minority of oral carcinomas. HPV (and also Epstein–Barr

virus) has also been blamed to cause hairy leukoplakia on the tongue in patients infected with human immunodeficiency virus.

5. *Cancer*   HPV has been blamed to cause invasive cancers of the skin, larynx and genital tract, mainly of cervix and also of penis, vulva, vagina, perineum and anus.

Laboratory diagnosis   It can be carried out by

1. *Histopathology and cytopathology*   These reveal characteristic features of HPV infection.

2. *Electron microscopy*   Papillomavirus particles can be readily seen by electron microscopy in most warts but are less in number in genital warts.

3. *Immunocytochemistry*   The HPV capsid antigen in paraffin sections of tissues or in cell smears can be detected by an immunoperoxidase test using a commercially available antiserum prepared by immunization of rabbits with bovine papillomavirus particles disrupted with sodium dodecyl sulphate. It detects all the genital HPV types. However, monoclonal antibodies that can differentiate the genital types have not been produced and may be used as diagnostic tools.

4. *Detection of viral nucleic acid*   Viral DNA in fresh tissues and exfoliated cells can be detected by DNA hybridization and polymerase chain reaction.

POLYOMAVIRUSES

Polyomavirus (poly-many; oma-tumour) virions are non-enveloped, 42–45 nm in diameter with a 72 capsomer icosahedral capsid. Viral genome is a double-stranded supercoiled loop of DNA molecule with a molecular weight of 3.4×10^6 Da. Polyomaviruses include Mouse polyomavirus infecting rodents, Simian virus 40 (SV40) infecting monkeys, and JC and BK polyomavirus infecting humans.

JC polyomavirus was first isolated from the brain of a

male patient with Hodgkin's disease who developed progressive multifocal leucoencephalopathy (PML). It was designated by the initials of the person from whom it was isolated. The virus persists for life in the kidneys and is shed in urine sporadically and more frequently during pregnancy and immunosuppression. PML is a rare, fatal, subacute demyelinating disease of nervous system that results from JC polyomavirus infection of oligodendrocytes in the brain. It is seen mainly as a complication of advanced disseminated malignant conditions such as Hodgkin's disease or chronic lymphocytic leukaemia and primary or secondary immunodeficiency syndromes, especially AIDS or following immunosuppression for organ transplantation.

BK polyomavirus was isolated from the urine of a renal transplant recipient and was named after his initials. It causes subclinical infection of children before the age of 10. It may, however, cause upper respiratory symptoms. The virus persists for life in kidneys. Reactivation may occur during last trimester of pregnancy and following immunosuppression for organ transplantation leading to asymptomatic shedding of virus in urine.

Replication

Entry of virus through receptor (sialic acid) mediated endocytosis; transported to the nucleus, and uncoating of viral DNA occur without participation of lysozomes. Replication involves two stages, with early viral protein being read off of one DNA strand in counterclockwise direction and late viral proteins being read off the other strand in clockwise direction. Transcription is mediated by cellular RNA polymerase II, which results in early viral protein (large 'T' and small 't' antigens) these promote synthesis of late proteins VP1, VP2 and VP3. 'T' antigen is involved in the initiation of DNA replication. Assembly of virions occurs in cell nucleus. Viral particles are released by lysis of cell.

Laboratory Diagnosis

1. **Electron microscope**   Human polyomaviruses can be detected by electron microscopy of diseased brain tissue from cases of PML (JC polyomavirus) and in the urine of a renal allograft recipient (BK polyomavirus).

2. *Cytopathology*   Cytological examination of exfoliated urinary epithelial cells shows the presence of enlarged deeply staining inclusion.

3. *Virus isolation*   BK polyomavirus can be isolated from urine in human diploid fibroblast and JC polyomavirus from urine or brain in human foetal glial cell culture. These two virus can be differentiated by haemagglutination inhibition.

4. *Detection of viral antigen*   Viral antigen in urine can be detected by ELISA. Following brain biopsy or autopsy, JC polyomavirus antigen can be detected by immunofluorescence.

5. *Detection of viral nucleic acid*   It can be carried out by nucleic acid hybridization and PCR. Following brain biopsy or autopsy, JC polyomavirus DNA can be demonstrated by *in situ* hybridization.

HEPADNAVIRIDAE

The family *Hepadnaviridae* contains 5 hepatotropic viruses specific for man (HBV), woodchuck (WHV), ground squirrel (GSHV), duck (DHBV) and heron (HHV). These viruses contain dsDNA genomes and include persistent infections in their natural hosts. Only HBV causes human infections that will be discussed below.

Hepatitis B Virus (HBV)

HBV or Dane particle is a complex 42 nm double-shelled particle (figure 4.13a and c). Outer surface contain surface antigen (HBsAg) made up of lipid, protein and carbohydrate.

It encloses an inner icosahedral 27 nm nucleocapsid core which contain antigen called HBcAg. Inside the core is the genome of 3.2 kbp molecule of circular dsDNA, and a DNA–dependent DNA polymerase. During active viral replication hepatitis Be antigen (HBeAg) is derived from HBcAg and found free in the plasma.

Along with mature virion two other forms of HBV are found during infection: (1) spherical particles of 22 nm diameter and (2) elongated tubules of similar diameter, both of which are composed of only HBsAg (figure 4.13b). They normally occur in large numbers over the mature virions (100 to 1000-fold). HBsAg carries a group specific antigen 'a' and two type specific antigens, 'd' or 'y' and 'w' or 'r'. This gives rise to four antigenic types of HbsAg–adw, adr, 'ayw' and ayr. Type 'ayw' is predominant among parenteral drug users in places of Africa, Russia and India. Type 'adw' is predominant among homosexuals in regions of Europe and USA, and type 'adr' is predominant in Asia. The 'w' antigen has additional three variants such as 'q', 'x' and 'g'.

Properties   Virus gets inactivated by heating at 60°C for 10 minutes and by treatment with hypochlorite and 2% glutaraldehyde for 10 minutes.

Virion structure   HBV virions are double shelled particles whose outer lipoprotein envelope contains multiple related envelope glycoproteins. The most abundant protein on the virions surface is the 24 kDa hepatitis B surface antigen (HBsAg) or S protein, which is in fact identical to the Australia antigen. Two other important but less abundant proteins are also present in the envelope. The coding region for the S protein is located in a larger ORF that contains two upstream and in-phase AUG start codons. The region upstream of the S gene is referred as pre-S region, which is divided into pre-S1 and preS2 based on AUG presence (figure 4.14). Initiation of translation at the first AUG gives rise to the L or pre-S1 protein (39 kDa), involved in receptor binding, initiation at the second AUG produces the M or preS2 protein (31 kDa).

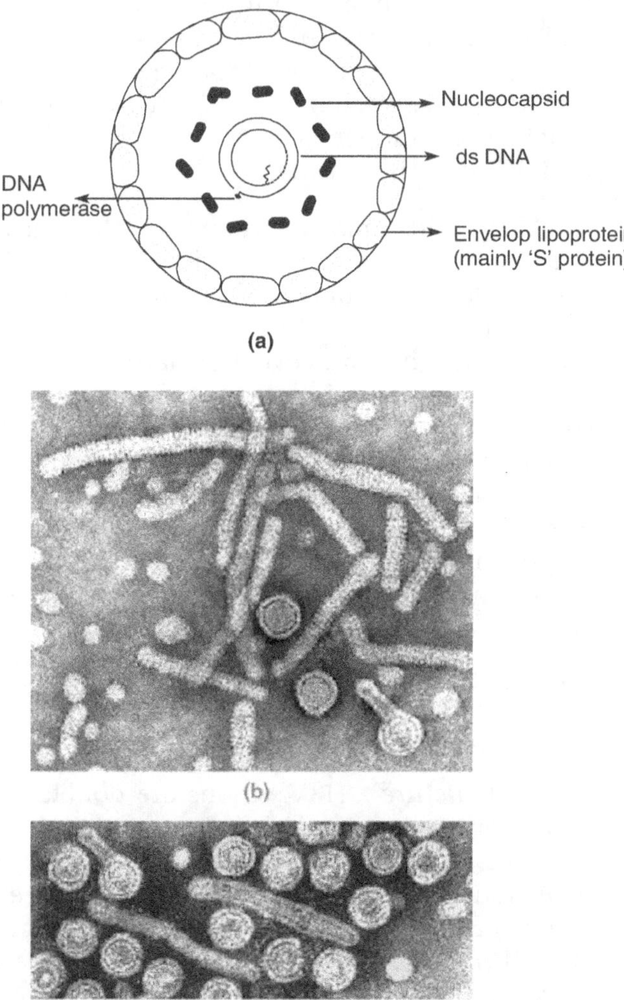

Figure 4.13 (a) Schematic diagram of Dane particle (42nm) or HBV virion which has lipoprotein envelope surrounding the nucleocapsid core containing dsDNA genome. One strand of DNA (+ strand) is less than unit length to that of the other strand (-strand) which is of 3.2 kb. (b) Electron micrograph showing elongated tubules of HBs Ag (c) Hepatitis B virus particles observed under electron microscope.

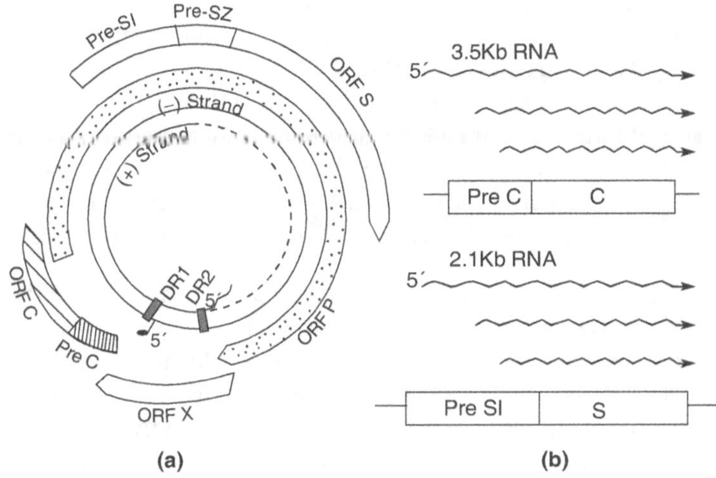

Figure 4.14 (a) The HBV genome: innermost circle depicts virion DNA. On this genome, filled circle indicate terminal protein on (-)strand, and wavy line denotes terminal RNA on (+)strand DNA. DR1 and DR2 indicate positions of an 11-nt. element directly repeated twice. Boxes denotes open reading frames (ORFs). (b) The fine structure of the 5′ termini of C/preCRNA, and the S/pre-S2 RNA.

The envelope of the virion can be removed by treatment with non-ionic detergents to liberate the inner nucleocapsid or core. This 27 nm structure contains the hepatitis B core antigen (HBcAg). The C protein is a 21 kDa polypeptide composed of two domains, an N-terminal domain that promotes oligomerization into capsids, and a C-terminal domain involved in sequence-nonspecific nucleic acid interactions.

Within the core is the viral genome, a relaxed circular molecule. One strand of viral DNA (minus strand) is unit length (3.2 kb), its complementary (plus strand) is less than unit length. The 5′ end of the (+) strand is fixed, but its 3′ end can be anywhere along the genome, which indicate the presence of a ss gap whose length is variable. The 5′ end of each viral DNA is covalently linked to another structure. Minus strands are linked to virus-encoded protein, (+) strands are linked to oligoribonucleotide. The termini of the (–) and (+) strands map to the regions of short (11–12 nt) sequences that are directly repeated twice on the viral DNA circle; these direct repeats are known as DR1 and DR2 as depicted in figure 4.14. They play an important role during viral DNA synthesis. Within the core is also a DNA polymerase activity a product of gene P (for polymerase), which is linked to (–) strand.

Subviral particles is one of the signature feature of hepadnavirus infected cells, two distinct subviral lipoprotein particles 20 nm spheres and longer filamentous forms are observed. These contain only envelope glycoproteins and the host derived lipids. The spheres contain primarily the S protein; filaments, which are less abundant have large quantities of L. Both types lack C and P proteins contain no viral DNA and are non-infectious. The probable function of these particles is believed to adsorb the circulating neutralizing antibodies and thus shield virions from the immune response.

HBV Genome Organization

Genome organization is depicted in figure 4.14. HBV contains four long ORFs: pre-S/S, encoding the envelope proteins; C, encoding the core protein; P, codes for viral polymerase; and X, encodes a small (17 kDa regulatory protein that influences the expression of viral and cellular genes.

Within the viral C gene, two in-frame AUG codons exist: the internal AUG encodes the 21 kDa C protein, the structural polypeptide of the viral capsid, whereas the most 5´ AUG directs production of the 24 kDa pre-C protein, wherein the pre-C region encodes a signal sequence that directs the protein into endoplasmic reticulum, and are removed via cleavage by signal peptidase. The remaining polypeptide is then secreted extracellularly, during this process the C-terminal region of the chain is removed by host proteases to generate a 16 kDa fragment known as HBeAg. HBeAg plays no role in virus assembly and its true in vivo function is not yet clear, however, the presence of HBeAg in the peripheral blood is correlated with a high level of viraemia and correspondingly high infectivity. Thus, a high level of HBeAg production is a marker for high levels of genomic RNA synthesis and hence of vigorous viral genome replication.

HBV Replication–An Overview

Figure 4.15 summarizes the main steps involved in replication cycle of HBV. The prime feature is that the DNA genome is replicated by reverse transcription of an RNA intermediate. The steps are stated below:

- Membrane fusion of the virus is followed by release of nucleocapsid into the cytosol.
- Transportation of viral core into the nucleus.
- Repair of DNA genome into covalently closed circular (ccc) form, which serve as transcription template for host RNA polymerase II.
- Generation of series of genomic and subgenomic transcripts employing an episomal template model.
- Transportation of all viral RNA into the cytoplasm, where translation yields the viral envelope, core, and polymerase proteins, as well as the X and pre-C proteins.
- Assembly of nucleocapsids takes place in the cytoplasm, during which a single molecule of genomic RNA is selectively incorporated by site specific binding of P protein.
- Once the P–RNA complex is formed, RNA packaging and reverse transcription begin. Virtually all DNA synthesis proceeds within the core rather than in free cytoplasm.

Completed particle bud into regions of intracellular membranes bearing the viral envelope proteins, wherein they acquire viral L, M, and S proteins.

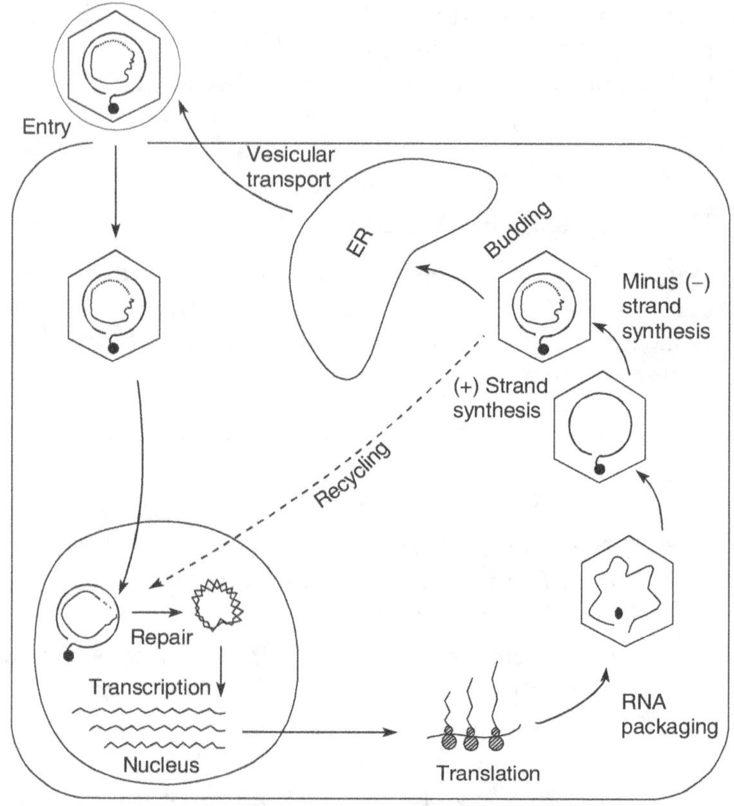

Figure 4.15   An overview of HBV replication cycle.
(ER - Endoplasmic reticulum)

Modes of transmission   These viruses are transmitted by parenteral, perinatal and sexual modes (Table 4.8, 4.8a).

1.   *Parenteral transmission*

HBV is present in blood and in body fluids such as semen, vaginal secretions, saliva, colostrums and breast milk. The concentration of HBV is much higher and that 1 ml of blood contaminating a syringe or needle, can transmit hepatitis B from one individual to another. Needlestick injuries, use of contaminated needles and syringes, intravenous drug (IVD) abuse, tattooing, acupuncture, sharing of razor and kissing can readily transmit HBV infection.

2.   Perinatal transmission

HBV can be transmitted from carrier mothers to their babies during the perinatal period. Transmission probably occurs when maternal blood contaminates the mucous membranes of the new born during birth. Transplacental transmission is thought to be quite rare, and breast feeding has never been observed as a mode of transmission.

3.   Sexual transmission

Since HBV is present in body fluids it can be easily transmitted by sexual contact. Homosexuals are particularly at higher risk. Most of the HBV infections are subclinical during the childhood. The course of clinical illness can be grouped into three phases viz., *Preicteric* or prodromal, *Icteric* or symptomatic, and *convalescent* or recovery. In the preicteric phase after an incubation period of 6 weeks to 6 months, patient develops malaise, anorexia, weakness, myalgia, nausea, vomiting and pain in the right upper abdominal quadrant. A small proportion of the infected area develops arthralgia, urticarial or maculopapular rash, polyarteritis nodosa and glomerulonephritis.

Icteric phase starts 2 days to 2 weeks following the initial symptoms wherein patients develop jaundice, pale stools and dark urine (bilirubinuria). Biochemical detection of hepatocellular damage is possible before the onset of jaundice and persists after it has resolved.

Convalescent phase is long and include malaise and fatigue lasting for several weeks after HBV infection is resolved, mild symptoms may persist for more than one year. The carrier rate of HBV in India is estimated to be 5%. Mild cases that do not result in jaundice are termed anicteric. Less than 1% of the icteric cases die of fulminant hepatitis, 90–95% recover with complete regeneration of the damaged liver within 2–3 months.

Remaining patients progress to chronic active hepatitis, cirrhosis and hepatocellular carcinoma (HCC). Hepatoma cells often contain HBV DNA, but the patient is usually negative for HBcAg and other indications of ongoing viral replication. About 10% of HBV infections result in chronic carrier state (i.e., persistence of HBsAg in circulation for more than 6 months). Carriers again are of two types: Super carriers who have high titers of HbeAg, a marker of active replication, and Simple carriers who have low levels of HBsAg and absence of HBeAg in their blood.

Laboratory diagnosis   In acute viral hepatitis serum transaminases (aminotransferases) are increased from 5 to 100-fold. Both alanine aminotransferase (ALT) and aspartate aminotransferase (AST) levels rise together late in the incubation period. Peak level is achieved at the time of jaundice and reverts back to normal in next 2 months. Serum bilirubin levels may rise up to 25-fold. Serological markers of HBV diagnosis is

Table 4.7   Serological markers of hepatitis B virus infection.

Clinical condition	Serological markers span						
	HBs Ag	*Hbe Ag*	*Anti– HBs*	*Anti– HBe*	*Anti–HBc*		*HBV DNA*
					IgM	*IgG*	
Incubation period	+	+	–	–	–	–	+
Acute hepatitis	+	+	–	–	+	+	+
Chronic active hepatitis	+	+	–	–	+	+	+
Asymptomatic carrier	+	–	–	+	–	+	–
Past infection	–	–	+	–	–	+	–
Immunized individual	–	–	+	–	–	–	–

stated in Table 4.7, which can be done by using sensitive and specific ELISAs.

HBsAg can be detected in the patient serum during incubation period, acute hepatitis, chronic active hepatitis and asymptomatic carrier state. HBeAg can be detected in the first three conditions and absent in asymptomatic cases, while anti-HBe indicate asymptomatic carrier state of the individual. Anti-HBcIgM indicates acute and chronic active hepatitis, while anti-HBcIgG in addition is present in asymptomatic carrier and past infection states. Viral DNA is detected only during incubation period and active hepatitis.

Prophylaxis   Measures for the control of HBV infection include screening of blood donors, use of sterile disposable syringes and needles, maintaining single sexual partner, use of condoms, and strict clinical practices like wearing gloves, gowns, masks, eye glasses to prevent exposure to blood and body fluids, and cleaning of blood spills with 2% glutaraldehyde or 0.5% sodium hypochlorite. Disposable equipments should be properly incinerated and others thoroughly sterilized.

Passive immunization   Hepatitis B immunoglobulin (HBIG) is prepared from donors with high titers of anti-HBs. This is given in doses of 300–500 IU intramuscularly (i.m.) to individuals after accidental exposure of blood from HBsAg positive patient within 48 hrs and, also to newborn infants born to HBsAg positive mothers.

Active immunization   The vaccines that are available for active immunization for children, health care workers, medical personnels, and high risk group of individuals (parenteral drug users, sex workers, patients on dialysis, transfusion and haemophiliacs) are discussed below.

1.   *Plasma-derived vaccine*

It consists of purified 22 nm particles of HBsAg, prepared from the plasma of symptomless carriers. The particles are separated by ultracentrifugation and treated with proteinase, 8M urea and formaldehyde. The chemical treatment inactivates

HBV, HIV and other contaminating viruses. The product is immunogenic and safe. This vaccine is still being produced and used, particularly in developing countries where it is more needed. The standard regimen was 3 doses of vaccine (0, 1, and 6 months) at doses of 20 μg in adults and 10 μg in children.

2. *Recombinant yeast hepatitis B vaccine*

It is produced by cloning the HBsAg gene in *Saccharomyces cerevisiae* yeast and the HBsAg particles are produced, extracted and purified for use as vaccine. This vaccine is safe, antigenic, free from side effects and as immunogenic as the previous one. Vaccine is adsorbed with aluminium hydroxide as adjuvant, stored in cold but not frozen condition. This is given as i.m. injection at 0, 1, and 6 months covering one course. Available as Engerix-B (Smithkline Beecham) and Recombivax HB (Merck Sharp and Dohme).

3. *Synthetic peptide vaccine*

These are chemically synthesized polypeptide vaccines that are safe and cheap.

4. *Hybrid virus vaccine*

Potential live vaccines using recombinant vaccinia virus have been prepared for hepatitis B, influenza, rabies, EBV and HIV. Recombinant vaccines can be generated by incorporating foreign genes (i.e., DNA sequence coding for HBsAg) into vaccinia virus DNA. Animal cells are first infected with vaccinia virus. Subsequently, a plasmid containing the foreign gene of interest, and promoter and thymidine kinase sequences from vaccinia virus is introduced into the cell. During replication of vaccinia virus DNA, the plasmid sequences are also replicated and chimeric viral DNA containing the foreign gene is produced. Recombinant vaccinia virus expresses proteins (HBsAg) encoded by foreign gene. It is cost effective, long shelf–life and possible use of polyvalent antigens.

5. *Therapeutic vaccines*

A recently developed therapeutic vaccine, called Theradigm–HBV (Cytel, SanDiego, CA) is capable of inducing an

HBV-specific MHC class I restricted (TL response in healthy volunteers. The vaccine consists of 3 components: HBcAg peptide, a T-helper peptide (tetanus toxoid peptide), and two palmitic acid molecules as lipid component. This vaccine is under phase II/I clinical trial.

6.   Antivirals

Nucleoside analogs such as lamuvidine, famciclovir and lobucovir are promising new agents currently in clinical trials.

Hepatitis D Virus (HDV)

HDV belongs to the genus Deltavirus and is a defective satellite virus requiring HBV as helper virus. It is spherical measuring 36–38 nm in diameter with HBsAg coat protein and HDAg nucleoprotein. The genome consists of a single-stranded circular minus sense (–)-RNA of 1.7 kbp. It encodes its own nucleoprotein. The outer coat of HDV virion is HBsAg encoded by the genome of HBV co-infection in the same cell. Replication of HDV requires the concomitant expression of HBV gene products.

Pathogenesis   Pathogenesis of HDV include two types of infections, one is a simultaneous co-infection with HBV commonly resulting from parenteral transmission as seen in intravenous drug users (IDUs). Secondly, super infection of an HBsAg carrier by HDV. It is more common and serious as it promotes HDV replication instantly without any delay. It leads to severe liver damage, fulminant HBsAg positive cases and increased mortality ~20%.

Clinical manifestation   The clinical features of acute HDV infection acquired through co-infection varies in serological profiles and is recognized only from the delayed raising of IgM and IgG anti-HDV. In severe disease viraemia is shown by early defection of HDAg and HDV-RNA in serum followed by sero conversion first to IgM and then to IgG anti-HD. Both types of response wanes within few weeks to months.

In superinfection IgG antibody persists for 1–2 years both antibodies rise to high titers and persist indefinitely that progress to chronicity. HDV-RNA can be demonstrated in blood.

Laboratory diagnosis   This includes detection of HBsAg and HDAg in the serum by ELISA, HDV RNA detection by northern hybridization, in later stages of acute disease, anti-HD IgM appears followed by anti-HD-IgG detected by ELISA. The clinical and epidemiological features of different hepatitis viruses are stated in Tables 4.8 and 4.9.

Table 4.8   Epidemiological characteristics of human hepatitis viruses.

Property	HAV	HBV	HCV	HDV	HEV
Genome	RNA	DNA	RNA	RNA	RNA
Virus family/ Genus	*Picorna- viridae*	*Hepadna- viridae*	*Flavi- viridae*	*Delta- virus*	*Calci- viridae*
Mode of trans- mission	Enteric	Parenteral, perinatal, sexual	Parenteral, sexual	Parenteral	Enteric
Antigen in blood	HAV	HBsAg, HBeAg,	HCV	HDAg	HEV
Antibody in blood	α–HAV	α–HBs, α–HBe, α–HBc	α–HCV	α–HDV	α–HEV
Chronic carrier state	No	5–10%	50%	>50%	No
Chronic hepatitis, cirrhosis	No	1–5%	20%	>50%	No
Liver cancer	No	Yes	Yes	No	No

Table 4.9   Viral hepatitis clinical and epidemiological features.

Features	HAV	HBV	HCV	HDV	HEV
Incubation (days)	15–45, M = 30	30–180, M = 60–90	15–160, M = 50	30–180, M = 60–90	14–60, M = 40
Onset	Acute	Acute or insidious	Insidious	Acute or insidious	Acute
Transmission:					
Faecal-oral	+++	–	–	–	+++
Percutaneous	–	+++	+++	+++	–
Perinatal	–	+++	±	+	–
Sexual	–	++	±	++	–
Progression to chronicity (%)	0	1–10 (90% of neonates)	70–90	Common invariable in HDV superinfection	0
% with fulminant hepatitis	0.1	0.1–1	Very rare	5–20	1–2 (10–20% in pregnant cases)
Prognosis	Excellent	Worse with age 7 debility	Variable	Acute—good, chronic—poor	Good

PICORNAVIRIDAE

Family *Picornaviridae* comprises of a large number of very small RNA (*Pico* = small, *rna* = RNA viruses) with a diameter of 27–30 nm. The capsid is a naked icosahedron made up of 60 protein subunits (protomers). The genome consists of a single linear molecule of single-stranded RNA of positive polarity with 7–8 kbp. This family is divided into five genera, three of which, Enterovirus, Rhinoviruses and Hepatovirus possess human pathogens that cause diseases (Table 4.10) other two genera, Aphthovirus and Cardiovirus cause foot-and–mouth disease and meningo-encephalomyelitis in mice respectively.

Table 4.10   Syndromes caused by enteroviruses other than polyviruses.

Genus	Species		Major Disease
Enteroviruses	1.	Polioviruses 1–3	Paralytic poliomyelitis, aseptic meningitis
	2.	Coxsackieviruses A1–24 except 23	Aseptic meningitis, encephalitis, herpangina, conjunctivitis (A24)
	3.	Coxsackieviruses B1-6	Encephalitis, fatal neonatal disease, pleurodynia, myocarditis or pericarditis
	4.	Echoviruses 1–34 except 10 and 28	Aspetic meningitis, encephalitis, rashes febrile illness, GI tract
	5.	Enteroviruses 68–71	Conjunctivitis (Enterovirus 70), polio-like illness and Hand, foot-and-mouth disease (enteroviruses71), pneumonia and bronchitis
Rhinovirus	Rhinoviruses 1–100		Common cold
Hepatovirus	Hepatitis A Virus		Hepatitis

Enteroviruses

Enteroviruses are among the most stable viruses. They can remain viable for years at –20°C to –70°C and for months at

4°C. In faeces at room temperature, the virus can remain infective for several weeks. Being non-enveloped they are inactive to ether, chloroform and deoxycholate. The virus is readily killed by moist heat at 50–55°C, but in food stuffs the virus may survive exposure to 60°C and holder method of pasteurization. Enteroviruses are rapidly inactivated by ultraviolet light, drying, formaldehyde (0.3%), hydrochloric acid (0.1 M) or free residual chlorine (0.3–0.5 ppm). However, higher concentrations of chlorine are necessary to inactive virus in the presence of organic matter because the latter diminishes the activity of chlorine.

Polioviruses

Poliomyelitis was a very dangerous and disabling disease. Although Landsteiner and Popper established the viral nature of poliomyelitis in 1909, the term *poliovirus* first appeared in 1955. Poliovirus was the first animal virus to be purified and obtained in crystalline form. They are isometric particles about 27 nm in diameter.

On the basis of neutralization tests, Polioviruses can be divided into three serotypes. Type 1 is the common epidemic type, type 2 is usually associated with endemic infections and type 3 occasionally causes epidemics. These viruses have affinity for nervous tissue and narrow host range. Only man and some primates like cynomolgous and rhesus monkeys are susceptible. These monkeys can be infected by the oral route and develop paralysis.

Natural infection occurs only in man. The virus is spread from man to man by faecal-oral route and because early multiplication occurs in both the oropharynx and the intestinal mucosa, the virus is also spread by pharyngeal secretions (droplet infection) during the first week of illness. No intermediate host is known.

On entering the body of the new host, the virus multiplies in the tonsils and Peyer's patches of the ileum. After multiplication, the virus enters the blood stream enabling the virus to become disseminated throughout the body including spinal cord and brain.

Predisposing factors

1. Neural spread may occur in children with inapparent infection at the time of tonsillectomy. Poliovirus present in the oropharynx may enter nerve fibres exposed during surgery and spread to brain resulting in bulbar paralysis.

2. A similar mechanism of viral spread via neural pathways may be responsible for paralysis of a limp recently injected with inflammatory injections such as diphtheria–pertussis–tetanus (DPT) vaccine. This is probably associated with the irritant properties of the adjuvant.

3. Pregnancy increases the incidence of paralysis.

4. Muscular activity during the paralytic phase of the illness may lead to paralysis of the limbs used.

Pathogenesis

Transmission occurs by faecal-oral route and virus gets shed up to 2 weeks in nasopharyngeal secretions and up to several weeks to months in faeces.

- Human cellular receptor for poliovirus maps to chromosome 19.

- Determination of neurotropism and neurovirulence for poliovirus have been investigated—viral RNA with 2 single base change leads to attenuation of previously neurovirulent poliovirus.

Clinical features

Inapparent infection   Patient does not have any symptom but the virus may be isolated from stool or throat or both. This is seen in 90–95% individuals.

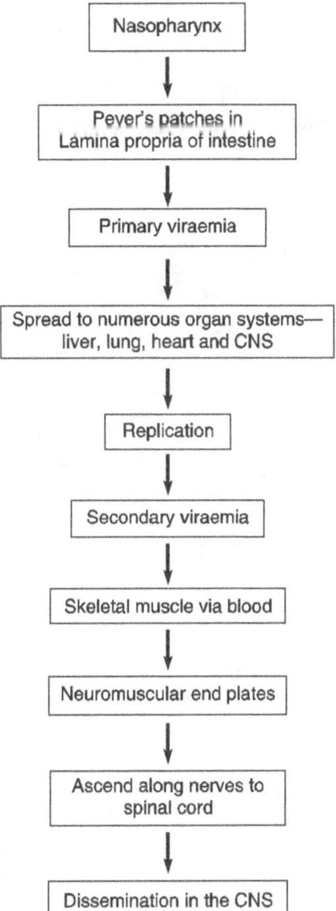

Minor illness (abortive type)   Patient develops mild, transient "influenza-like" illness along with anorexia, myalgia, fatigue, fever and sore throat. This is seen in 4–8% cases.

Non-paralytic poliomyelitis   In addition to "influenza-like" illness, patient also develops headache, neck stiffness, and back pain that may indicate some degree of aseptic meningitis. The illness lasts 2–10 days with rapid and complete recovery. This is seen in 1–2% individuals. Almost 10% of patients with abortive poliomyelitis develop aseptic meningitis (Table 4.10).

Paralytic poliomyelitis   Patient develops paralysis during the course of the illness. It is uncommon occurring in 0.1–2% poliovirus infections. On the basis of the site of involvement, the paralysis may be spinal, bulbar or bulbospinal.

Laboratory diagnosis

- Virus isolation from faeces and throat swabs taken early in disease. Cultivation done on monkey kidney, human amnion, HeLa, HEp2 BGM, MRC-5 and other cell cultures—cytopatheic effect seen within 48 h with cell retraction, increased refractivity, cytoplasmic granularity and nuclear pyknosis.
- Serotypic determination by neutralization tests.
- Differentiation of wild and attenuated vaccine strain by virulence tests in the monkeys and nucleic acid hybridization assays.

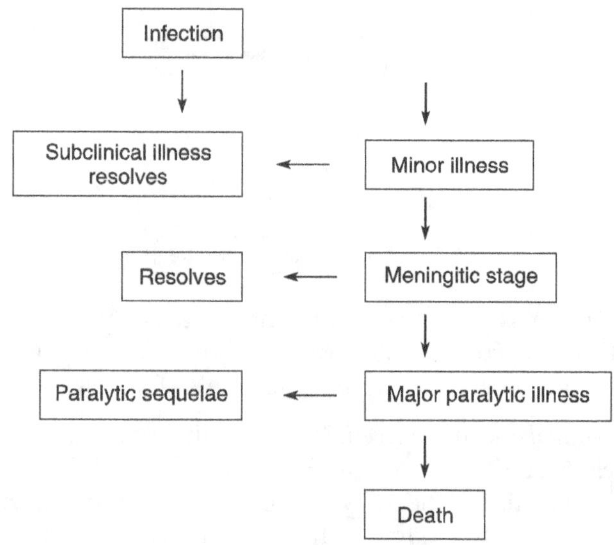

Prophylaxis

Inactivated polio vaccine (IPV)—Salk vaccine   Developed by Jonas Salk in 1956 for immunization by parenteral injection. The vaccine contains formalin inactivated strains of the three serotypes of the virus grown in monkey kidney cell culture, administered by deep subcutaneous or intramuscular injection. Three injections are given with intervals of 6–8 weeks between the first and second doses and 4–6 months between the second and third doses. It stimulates production of serum IgG with low levels of secretory IgA. Thus it can prevent central nervous system (CNS) involvement of the virus but not alimentary tract infection. The absence of live virus makes it safe to administer to immunocompromised individuals.

Live attenuated oral polio vaccine (OPV)—Sabin vaccine   Developed by Albert Sabin in the year 1962. It contains live attenuated strains of the three serotypes of poliovirus grown either in cultures of monkey kidney cells or human diploid cells. The virus is unable to multiply in the CNS and lacks neurovirulence. The vaccine is administered orally and simulates natural infection. It induces both local secretory IgA antibodies in the pharynx and alimentary tract , and humoral IgG antibodies in circulation. Vaccine virus is excreted in the faeces for several weeks post vaccination, during which period virus spread may occur to close contacts, inducing or boostering immunity in them. Dose regimen includes: The risk of vaccine-associated poliomyelitis has been estimated at between 0.5 and 3.4 cases per million.

Coxsackieviruses

Viruses are named after the first isolation from Coxsackie village in New York by Dalldrof and Sickles (1948). Based on the pathological changes produced in suckling mice, they are classified into two groups 'A' and 'B'. Group A viruses produced generalized myositis, flaccid paralysis followed by death within a week in suckling mice. Number of serotypes includes 23. Group B viruses form patchy focal myositis, spastic paralysis and localized lesions in the liver, pancreas, myocardium, brain and brown fat pads. There are 6 serotypes under group B.

Pathogenesis   Spread by faecal-oral route and cause the following clinical features:

Group A Viruses cause Aseptic meningitis (by serotypes 2, 4, 7 and 9), herpangina (vesicular pharyngitis) caused by serotypes 2, 4, 5, 6, 8 and 10. The highest incidence of herpangina is among children 1 to 7 years old, usually has an abrupt onset of fever associated with sore throat, dysphagia and malaise. Early in the illness, grayish-white discrete vesicles (1-4 mm diameter) appear over palate, and tonsillar pillars and on the oropharynx. The vesicles rupture, leaving a punched-out ulcer. Foot-and-mouth disease caused by types 5 and 16 predominant in childhood stages.

Group B *viruses* cause epidemic myalgia, a febrile illness with pain in the muscles of the chest and epigastrium, myocarditis and pericarditis, and aseptic meningitis.

Laboratory diagnosis

1.   Inoculation by intracerebral and or intraperitoneal routes of newborn mice.
2.   Growth in human diploid embryonic lung fibroblasts, BGM, Hela cell line—cytopatheic effect (CPE) resemble those of poliovirus.
3.   Neutralization test with reference antisera.

ECHO Virus

These viruses are preliminarily isolated from the faeces of individuals who had no clinical illness and caused a CPE in cell culture, thus given the name of enteric cytopathogenic human orphan viruses (ECHO). They are classified into 32 serotypes on the basis of their type-specific neutralization capsid antigen. Most of them produce asymptomatic infections but some of them have been associated with aseptic meningitis, paralysis, fever with rash, respiratory diseases, infantile diarrhoea, pericarditis and myocarditis. Diagnosed by isolation in human diploid embryonic lung fibroblast cultures from nose, throat swabs, stools or CSF specimens.

Enteroviruses 68–71

Of the four, 68, 70 and 71 are associated with disease in humans. Type 68 causes pneumonia and bronchitis, type 70 with acute haemorrhagic conjunctivitis (AHC) and type 71 with meningoencephalitis and paralysis. AHC is highly infectious with an incubation period of about 24 hrs and the symptoms include sudden swelling, congestion, watering and pain in the eyes. Recovery was observed in 3–7 days. Isolation of the viruses are obtained in human diploid embryonic kidney and HeLa cell line.

Rhinoviruses

The name rhinovirus stems from the special adaptation of the virus to the nasal passages.

Rhinoviruses are small RNA viruses morphologically similar to other picornaviruses, but can be differentiated on the basis of their acid liability and optimal temperature of 33°C for replication. They are inactivated at low pH (<5.0), and can survive on environmental surfaces for several days. Few of them can withstand heat at 50°C for 1 hr. They are resistant to 20% ether and 5% chloroform but are sensitive to hypochlorites and aldehydes. They have been subdivided into more than 102 serotypes based on the type-specific antigen in their capsid. Rhinoviruses can be isolated in human embryonic kidney cells, nasal cells or tracheal epithelium. Some strains grow both in monkey and human cell lines and are designated as 'M' strains, and 'H' strains for those which multiply only in humans.

Composition   The virion is composed of a ssRNA genome (+sense) of 7.2 kb long packed into a protein capsid comprising 60 protein subunits (protomers) organized as 12 pentamers.

Water is an important part of the virion structure. Each of the 12 pentamers contains a prominent valley or "Canyon" forming a central plateau at base filled with hydrophobic molecules called pocket factor.

Replication　Multiplication is entirely cytoplasmic and completed virus particles are ultimately released by disintegration of the host cell.

Pathogenesis　Rhinoviruses are the major cause of common cold accounting for about half of all colds. Others implicated in common cold include coronaviruses, some enteroviruses such as coxsackieviruses A21 and A24, ECHO viruses 11 and 20, RSV and parainfluenza viruses. Rhinoviruses are transmitted by aerosols contamination expelled from the nose of the patient during sneezing and coughing. During the acute phase of the illness, high concentrations of the virus are present in nasal secretions which may contaminate hands, fingers, handkerchiefs, door knobs, etc. normal individuals who touch these contaminated surfaces acquire the virus. When they touch their eye or nasal mucosa leads to common cold.

Clinical features　Incubation period ranges from 2–4 days and patient develops profuse watery discharge (rhinorrhoea) with nasal obstruction, sneezing, sore throat, cough, headache, malaise with low grade fever. Symptoms subside normally in a week's time but may be prolonged for 2 weeks or longer. Sinusitis may supervene if secondary bacterial infection occurs. Recovery is mediated by endogenous interferons, locally synthesized IgA antibodies and serum IgG antibodies.

Laboratory diagnosis

1.   Isolation in human and monkey cell lines from nose, throat swabs or nasopharyngeal aspirates—incubation at 33°C.
2.   ELISA based detection of virus antigen from nasal washings.
3.   Detection of viral RNA extracted from nasal washings by RT–PCR.

Hepatitis A virus (HAV)

HAV, formerly enterovirus 72, has now been assigned its own genus on the basis of its distinctive nucleotide sequence. HAV belongs to the genus Hepatovirus of the family *Picornaviridae*. It is a non-enveloped, 27 nm icosahedral virus, having cubic symmetry. It comprises of linear ssRNA and has only one serotype. It is more stable and can withstand heating at 56°C for 30 minutes and treatment with diethylether. It can be grown in cell cultures of monkey and human cells.

Genome   The nucleic acid is ssRNA with positive polarity ~7.5 kb in length. It codes for 4 polypeptides, VP1, VP2, VP3 and VP4. The genome can be divided into 3 sections: (1) a 5´ non-coding region which is uncapped and has a viral protein VPg covalently linked at the 5' terminus; (2) an ORF coding all the viral proteins; (3) as short 3' non-coding region.

Replication   Genome multiplication occurs entirely in the cytoplasm, where the genomic RNA can act as mRNA directly, directing the synthesis of large viral polyprotein which is then cleaved into segments.

The initial step in the production of new viral RNA is to copy the incoming genome RNA to form complementary negative strand RNA. This serves as template for synthesis of positive strand RNAs.

Clinical features   The incubation period is 2–6 weeks. Illness starts with few days of malaise, loss of appetite, vague abdominal discomfort, and fever that comprise the prodromal (preicteric) stage. Hepatomegaly, due to cell necrosis, causes blockage of the biliary excretions resulting in jaundice. The urine then becomes dark and the faeces pale; soon afterwards jaundice becomes apparent, first in the sclera and then in the skin. The patient recovers in a week or so and the jaundice disappears in a months time. The fulminant form of hepatitis A and liver failure can occur in <0.5% cases. Complete recovery occurs in 2–3 months. Disease severity is more in adults with 50% than children <3 yrs with only 5%.

Pathogenesis   Virus shedding is observed 1–2 weeks prior to onset of symptoms, and persist for several days after the serum transaminase levels increases. There is hardly any virus in the serum which suggests that epidemiology of the disease as faeco-oral enteric infection. It probably multiplies first in the intestine and then viraemia occurs wherein it reaches the liver. Histopathology is similar to hepatitis B virus.

Laboratory diagnosis

1.   Marked rise in serum ALT and AST levels.
2.   Virus detection by immune electron microscopy.
3.   Faecal HAV antigen detection by ELISA.
4.   Detection of anti-HAV-IgM by ELISA.
5.   Isolation of virus from faeces in cell cultures of monkey kidney and human fibroblasts.
6.   Trace amounts of virus in food and water can be detected by RT-PCR.

Prophylaxis   Passive immunization with normal human immunoglobulin gives protection of seronegative individuals for a period of 4–6 months. Hepatitis A virus vaccine (formalin inactivated, adsorbed to alum as adjuvant) can be used for active immunization. Two doses of one month apart elicit a good immune response in 99% of the vaccines lasting for several years. It may be given to high risk individuals like long-term visitors to countries in which HAV is endemic, sewage workers, sexually active homosexuals and IDUs.

Hepatitis E Virus (HEV)

HEV is an acute and generally self limiting infection of the liver, but causes high mortality in pregnancy.

This virus belongs to the family *Caliciviridae* on the basis of its particle structure and overall genome organization. HEV is isolated from bile or faeces and are non-enveloped, icosahedral, ~32 nm in diameter.

Genome   Genome comprises of ssRNA of positive (+) polarity, to a length of 7277 nt in case of HEV US-2 strain including a 26 nt poly(A) tail. The 5´ end of genome has a 7-methylated Guanine cap. It has conserved 5´ and 3´ UTR region of 35 and 60 to 75 nt, respectively that play a role in replication and encapsidation. The genome contain 3 ORFs organized as 5´—ORF1-ORF3-ORF2-3´ with ORF 3 and 2 largely overlapping. These ORF encodes 3 proteins, viz. PORF1 a replicative polyprotein of 186 kDa, PORF3 with a possible regulatory protein of 13 kDa, and PORF2 a capsid protein of 660 amino acid residues. (figure 4. 16) The polyprotein codes for methyl transferase, papain like protease, helicase and RNA dependent RNA polymerase (RDRP) enzyme.

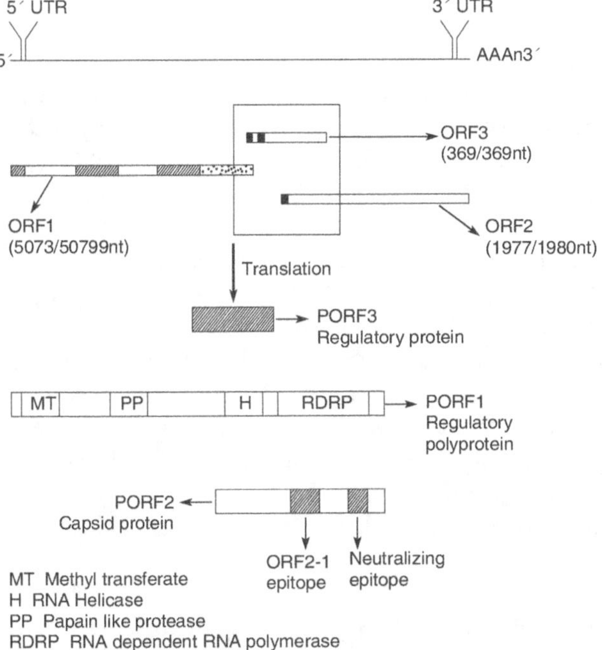

Figure 4.16   HEV genome structure and organization. Top panel shows positive sense single stranded (SS) polyadenylated (An) RNA genome flanked by untranslated regions (UTRs) of 27 nucleotide (nt) and 68 nt at 5´ and 3´ end respectively. Panel below shows three different coding regions (open reading frames-ORFs) of the virus and encoding proteins.

Replication   Following interaction of HEV with the cellular receptor, uncoating of virus occurs. The released viral RNA serves as mRNA (for ORF1), after cleavage of PORF1 polyprotein the replicative protein, RDRP copies viral genome to complementary negative (-) RNA which serve as template for (+) RNA synthesis.

Growth in cell culture   HEV has proven largely refractory to cell culture, but demonstrated in macaque hepatocytes.

Transmission   Transmission is by faecal-oral route. Consumption of contaminated water is the most common route. Atypical routes are suggested from US based studies that observed putative zoonotic HEV infection.

Pathogenesis   Incubation period (i.p.) of HEV infection increased with increasing dose of virus, normally between 5–6 weeks. If higher dose is present i.p. may be lowered down to 2 weeks after exposure.

Clinical features   HEV infection begins with a flu-like symptom lasting for 1–10 days, with fatigue, malaise, anorexia, nausea, vomiting and low-grade fever. Dark urine and pale coloured stools are followed by onset of clinical jaundice. Spleen enlargement was seen in 10–15% patients. Recovery was observed within 4–6 weeks, and mild hepatomegaly and increase in ALT levels seen during this period. Complication due to HEV is the case of fulminant hepatitis observed in less than 1% cases and up to 30% during third trimester of pregnancy.

Laboratory diagnosis   This includes isolation of the virus in cell culture, antigen detection by immune electron microscopy and ELISA, nucleic acid detection by RT-PCR of HEV-RNA, serology of HEV by detection of IgM and IgG antibodies to HEV antigens. The common serological markers of hepatitis viruses are tabulated as shown in Table 4.11.

Table 4.11   Common serological markers and interpretation.

Virus	Pattern	Status
HAV	IgM α–HAV$^+$	Acute infection
	IgG α–HAV$^+$	Remote infection
	HBsAg$^+$, IgM α–HBc$^+$	Acute infection
	HBsAg$^+$, α–HBs$^-$, IgM α–HBc$^+$	Acute infection
HBV	HBsAg$^+$, IgG α–HBc$^+$	Chronic infection
	HBsAg$^+$, IgG α–HBc$^+$, HBV-DNA$^+$, HBeAg$^-$	Chronic, 'e' antigen infection
	HBsAg$^+$, IgG α–HBc$^+$, HBV-DNA$^+$, HBeAg$^+$	Replicative chronic infection
	HBsAg$^+$, IgG α–HBc$^+$, HBV-DNA$^-$, HBeAg$^-$	Non-replicative chronic infection
	HBsAg$^-$, IgG α–HBc$^+$, α–HBs$^+$	Resolved
	α–HDV$^+$, HBsAg$^+$	HDV infection
HDV	α–HDV$^+$, IgM α–HBc$^+$	Co-infection
	α–HDV$^+$, IgM α–HBc$^-$	Superinfection
HCV	α–HCV$^+$ and or HCV-RNA$^+$	HCV infection

REOVIRIDAE

Rotavirus

Rotavirus diarrhoea is a major public health problem throughout the world causing more than 125 million cases of infantile diarrhoea and about 1 million deaths per year. Although, rotavirus infection occur with nearly equal frequency in both developed and developing countries, it leads to an estimated 600,000–870,000 deaths per year among children under 5 years of age in developing countries, Whereas only 75–150 deaths per year were found to occur in the United States. In India, 20–30 % hospitalized cases of diarrhoea are due to rotaviruses. In view of the significant morbidity and mortality associated with dehydrating diarrhoea due to rotavirus, there is an urgent need to develop rotavirus vaccine for use in early infancy.

History

Rotavirus was first discovered in 1973 by Bishop *et al.* and the association of the 70 nm rotavirus with severe diarrhoea in infants and young children was also found. The identification relied on direct visualization by thin section electron microscopy (EM) of duodenal biopsies, obtained from children with acute non-bacterial gastroenteritis.

It soon became apparent that the 70 nm particle (figure 4.17), subsequently designated rotavirus, was an important etiological agent of childhood diarrhoea in both developed and developing countries.

Figure 4.17 Rota virus.

Structure   The term rotavirus is derived from the Latin word, 'Rota' which means 'wheel', and was suggested because, the sharply defined circular outline of the outer capsid gives the appearance of the rim of a wheel placed on short spokes radiating from a wide hub.

The human rotaviruses (HRVs) have a distinctive morphologic appearance by negative stain electron microscopy (EM). Complete particles measure ~70 nm in diameter and have a distinctive triple-layered icosahedral protein capsid that contains the virus genome. The genome comprises of 11 segments of double stranded (ds) RNA which range in molecular weight from 2×10^5 to 2.2×10^6 Da with a size range of 0.6 to 3.3 kbp.

Characteristics of Rotaviruses

Structure	• Mature virus particles measure ~70 nm
	• Possess triple-layered icosahedral protein capsid
	• Non-enveloped (resistant to lipid solvents)
	• Capsid contains all enzymes for mRNA production
Antigenic Composition	• Inner capsid contains major antigen VP6
	• VP6 determines group (A to G) and subgroup (SGI, II, I + II, nonI, nonII) specificity
	• Outer capsid contains two major neutralization antigens, viral protein 7(VP7) and VP4.
	• VP7 and VP4 determines 'G' and 'P' serotypes
Genome	• Contains 11 segments of double stranded RNA (dsRNA)
	• Each RNA segment codes for at least one protein
	• Codes for six structural VP1, VP2, VP3, VP4, VP6, VP7 and five non-structural NSP1, NSP2, NSP3, NSP4, NSP5 proteins.
	• Gene segment 4 codes for neutralization antigen VP4, segment 9 codes for VP7 and segment 6 codes for major group antigen VP6
	• Capable of high genetic reassortment during dual infections
Replication	• Cultivation facilitated by proteolytic enzymes
	• Cytoplasmic replication
	• Unique morphogenesis involves transiently enveloped particles
	• Mature particles are non-enveloped
	• Virus normally released by lysis

Antigenic composition Rotavirus particles are made of six polypeptides organized in a triple-layered icosahedral protein capsid, which contains the viral genome. An outer layer comprising of viral proteins (VP) VP7 and VP4, a middle layer

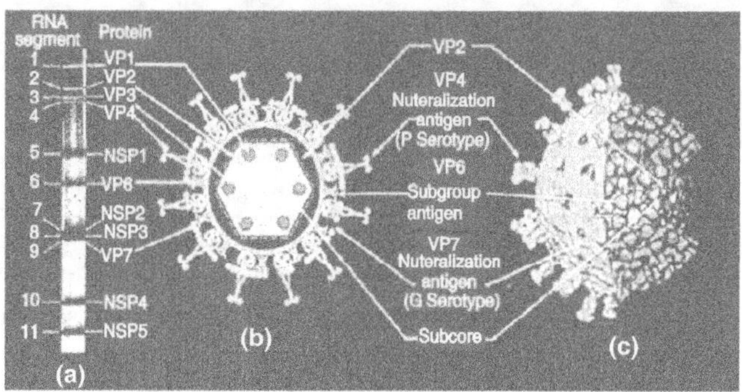

Figure 4.18   Gene coding assignments and the 3D structure of rotavirus particle. (a) PAGE showing 11 segments of dsRNA. (b) Schematic representation of complete rotavirus particle. (c) 3D structure of complete rotavirus particle (portion of outer shell and inner shell have been removed)— from Estes, 1996.

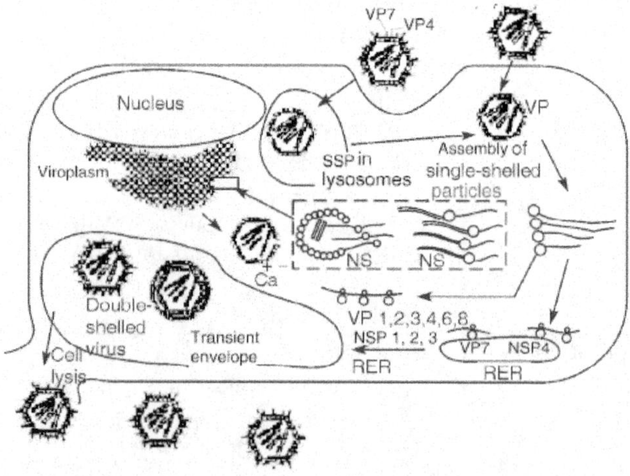

Figure 4.19   Overview of rotavirus replication.

of VP6 protein and an inner layer of VP2 protein which encapsidates the core containing two minor proteins VP1 and VP3, and the 11 segments of genomic dsRNA as shown in figure 4.18. An overview of the replication of rotaviruses is depicted in figure 4.19.

Rotaviruses possess three important antigenic specificities based on group, subgroup and serotype. Group specificity is determined by VP6 protein, the predominant group antigen. Subgroup specificity which serves as an important epidemiological marker is also mediated by the VP6 protein. The serotype antigenicity is associated with the two major outer capsid proteins VP7 and VP4. The schematic representation of major structural antigens of rotavirus is shown in figure 4.20.

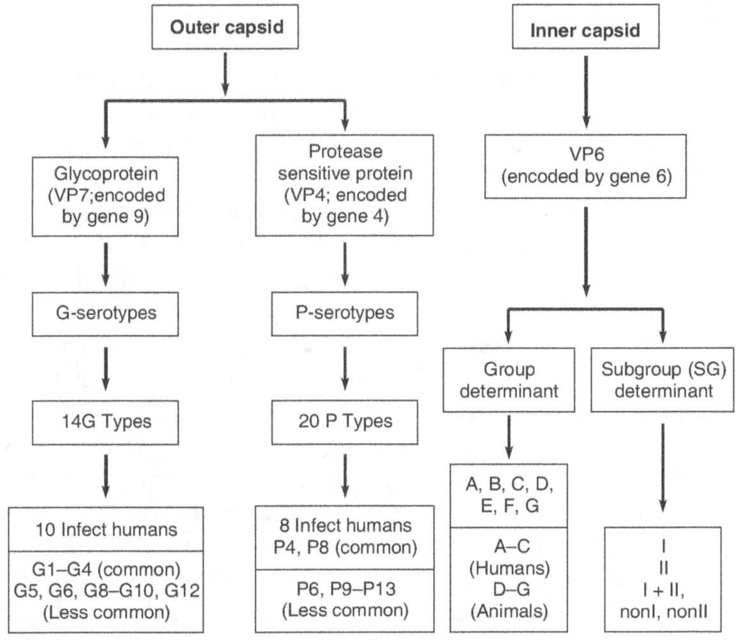

Figure 4.20 Schematic representation of major structural antigens of rotavirus.

Rotaviruses are subdivided into seven serogroups A to G, on the basis of their inner capsid group antigen VP6. Only rotaviruses in group A, B and C have been associated with disease in humans, and group A rotaviruses are the major cause of severe diarrhoea in infants and young children. Group A rotaviruses have been further classified into four subgroups (SG): SGI, SGII, SGI + II and nonI, nonII, depending on their reactivity to VP6 antigen which is encoded by gene segment 6. Within each group and subgroup, distinct serotypes are identified. Rotavirus serotypes have been defined according to two neutralization antigens, VP7 (a glycoprotein that determines 'G' serotypes) encoded by gene segment 7, 8 or 9 depending on the strain and VP4 (a protease sensitive protein that determines 'P' serotypes) encoded by gene segment 4, located on the outer capsid. To date, 14 rotavirus 'G' serotypes, including 10 that infect humans, and 20 'P' serotypes with 8 that infect humans have been identified by cross-neutralization studies.

Genome structure   The RNA segments fall into four size classes based on contour length measurements by EM and confirmed by nucleotide sequence analysis of the 11 genes. The distribution of the 11 segments into these four size classes is also evident by polyacrylamide gel electrophoresis (PAGE) of RNA. These RNA segments are numbered 1 to 11 in the order of migration in PAGE with the highest molecular weight segment designated as gene 1 and the lowest as gene 11. Group A rotaviruses is composed of four large segments (1 to 4), two medium sized segments (5 and 6), three smaller segments constituting a distinctive triplet (7 to 9) and the two smallest segments (10 and 11) showing the 4-2-3-2 pattern. In some HRV strains, the 11th (cognate) RNA segment migrates more slowly than usual, and is thus located between the 9th and 10th (cognate) segments yielding a "short" pattern characteristic of almost all antigenic subgroup (SG) I HRVs. Similarly, a faster migration of 10th and 11th RNA segment yielded a long pattern of HRVs which is characteristic of almost all SGII HRVs. In addition, strains with "super-short" RNA patterns have been recovered from humans, in which the 11th (cognate) gene segment migrated even more slowly than the short pattern. The gene coding assignments of rotavirus is shown in Table 4.12.

Table 4.12 Nucleotide sequences and gene coding assignments of group A rotaviruses.

Genome Segment	Protein Product[a]	Size range (bp)	Location in virus particles	Molecular weight (Da)[b]	Function
1	VP1	3302	Central core	125005	RNA polymerase
2	VP2	2687–2717	Central core	102431	RNA binding
3	VP3	2591	Central core	88751	Guanyltransferase
4	VP4 (VP5* + VP8*)	2359–2364	Outer capsid	86751	Haemagglutinin, protease-enhanced infectivity, virulence, cell attachment
5	NSP1 (NS53)	1578–1611	Non structural	58654	Metal binding domain
6	VP6	1356	Inner capsid	44816	Subgroup antigen hydrophobic, acidic,
7	NSP3 (NS34)	1075–1104	Non structural	34600	RNA binding

(Contd.)

Table 4.12   Contd.

Genome Segment	Protein Product[a]	Size range (bp)	Location in virus particles	Molecular weight (Da)[b]	Function
8	NSP2 (NS35)	1059	Nonstructural	36700	Basic, RNA binding
9	VP7	1062	Outercapsid	37368	Surface glycoprotein, neutralization antigen
10	NSP4 (NS28)	750–751	Non structural	20290	Transmembrane glycoprotein, role in morphogenesis
11	NSP5 (NS26)	663–667	Non structural	21725	Basic, RNA binding

(a) The virion polypeptides are designated as proposed by Mason *et al.* (1980) modified by Liu *et al.* (1988) and Mattion *et al.* (1994). The parantheses show the names of non structural proteins as designated previously. (b) Molecular weight are for simian rotavirus (SA11) proteins calculated from the deduced aminoacid sequences from the nucleotide sequence and from the longest potential open reading frame (ORF).

* On cleavage.

Epidemiology   Numerous studies on the distribution of group A rotavirus 'G' serotypes have demonstrated a high prevalence of serotypes G1 to G4 globally. However, unusual 'G' serotypes can be common in some parts of the world, especially developing countries e.g. G8 and G9 in India, and G5 and G10 in Brazil. Presently, rotavirus serotype can also be predicted by nucleotide sequence analysis of the VP7 gene and VP4 gene, because there is a high degree of conservation of sequence among rotaviruses belonging to the same serotype. The use of reverse transcriptase-polymerase chain reaction (RT-PCR) technique has enabled the genotyping of rotavirus specimens that could not be typed by ELISA. Since this method is based on genomic RNA, the term 'genotyping' is used in place of serotyping. This method is 1000 times more sensitive than ELISA and 10,000 times more sensitive than RNA-PAGE.

Molecular epidemiology   The term "electropherotyping" was applied to the method of differentiating strains based on migration patterns of RNA by PAGE. The existence of various electropherotypes (E-types) offer the possibility for studying molecular epidemiology of rotaviruses in different regions of the world. Molecular epidemiological studies have revealed the presence of numerous different electropherotypes which are broadly classified into "short", "long" and "super-short" RNA electropherotypes based on the relative migration rates of gene segments 10 and 11.

Studies on the distribution of electropherotypes showed, a sequential pattern of appearance and prevalence of different types. Characteristically in these studies, a single strain of rotavirus was predominant at any specific time, often accompanied by less common types.

Clinical features   Rotavirus infection produces a spectrum of responses that vary from subclinical infection to mild diarrhoea to a severe and occasionally fatal dehydrating illness. Three major clinical features observed in rotavirus gastroenteritis were vomiting, diarrhoea and dehydration. The incubation period in humans and most animals appeared to be 2–3 days. In a study conducted by Rodriguez *et al.* (1977), the clinical

Table 4.13   Viral agents that cause Gastroenteritis.

Virus	Morphology	Genome	Detection method	Comment
Rotavirus				
Group A	70 nm double capsid, wheel shaped ('rota'), 11-segments of dsRNA	Linear dsRNA, 18 kb, 11 segments	EIA, EM, PAGE, RT-PCR, culture	Common cause of severe diarrhoea, Traveller's diarrhoea, and immunocompromised patients
Group B				Found only in China
Group C				World-wide distribution, outbreaks in children in Japan, UK and US
Calicivirus	27–32nm, SRSV with ssRNA	Linear (+) ssRNA, 7.5 kb, has VPg protein at 5´ end and poly(A) tail at 3´ end	EM, IEM, RT-PCR, Southern-blot hybridization	Epidemic in all age groups, common pathogens of children
Astrovirus	27–32nm, ssRNA, SRV	Linear (+)ssRNA, 7kb, has poly(A) tail at 3´ terminus	EM, EIA, RT-PCRA	Children with <3 yrs are affected
Adenovirus	80–110nm serotype 40,41	Linear dsDNA, 36 kb, inverted terminal repeats and a protein primer at each 5´ end	EM, EIA	Endemic in children

manifestations of vomiting and severe dehydration was significantly higher among patients with rotavirus diarrhoea compared to those diarrhoeal illness which is not associated with rotavirus. Notable laboratory findings were related to moderate severe dehydration status of rotavirus illness (Table 4.13). Deaths have also been reported in infants and young children with rotaviruses infection in previous studies.

Other complication of rotavirus infection include intestinal haemorrhage and intussusception of small bowel. In neonates, rotavirus infection can be associated with necrotizing enterocolitis, persistent lactase deficiency or other carbohydrate malabsorption.

Assays for rotavirus detection   Many assays have been developed for the detection of rotaviruses in stools. Specimens from the first to fourth day of illness are ideal for rotavirus detection.

Initially, visualization of stool material by direct electron microscopy (DEM) was employed for rotavirus detection which permits detection of rotavirus in 80% to 90% of the virus positive specimens. DEM has the advantage of high specificity because rotaviruses have a distinctive morphologic appearance. DEM continues to be a mainstay in the diagnosis of rotavirus disease, and is frequently used as the final arbiter when discrepancies occur with other techniques.

Method	Efficiency[a]	Practicality[b]
Electron microscopy (EM)	4+	1+
Immune electron microscopy (IEM)	4+	1+
Complement fixation (CF)	1+	4+
Counter-immunoelectro-osmophoresis (CIEOP)	3+	4+
Immunofluorescence in cell culture and smears (IF)	1+	1+
Cell culture (CPE)	4+	2+ to 3+

(Contd.)

Method	Efficiency[a]	Practicality[b]
Radio immunoassay (RIA)	4+	3+ to 4+
Enzyme linked immunosorbent assay (ELISA)	4+	4+
Immune adherence hemagglutination assay (IAHA)	3+	2+ to 3+
RNA electrophoresis pattern in polyacrylamide gels (PAGE)	3+	3+

(a)   On a scale of 1+ to 4+, where 1+ indicates low degree of efficiency or practicality and 4+ indicates a high degree of efficiency or practicality.

(b) For large-scale epidemiological studies, assuming 4+ efficiency.

ELISA is the most widely used method for the diagnosis of HRV infections. The method has the advantage of screening samples on large scale rapidly. Presently, Mab ELISA has been used in the detection of rotavirus from faecal specimens.

Polyacrylamide gel electrophoresis (PAGE)   All rotaviruses contain a genome of 11 double-stranded RNA segments contained within the core of the virus. The genomic RNA segments can be resolved by PAGE and visualized by ethidium bromide or silver staining. The resulting RNA profiles are called genome pattern or electropherotypes (E-types). E-types have been widely used to characterize isolates from outbreaks.

PAGE and silver staining technique was performed previously by several methods. These include briefly the initial extraction of dsRNA from the virus, followed by electrophoresis on polyacrylamide gels.

RT-PCR   PCR amplification of rotavirus nucleic acid from stool specimens has been described and its sensitivity has been reported to be 1000 times higher than ELISA and 10,000 times higher than PAGE. Most PCR studies on group A rotaviruses are limited to characterization into G and P genotypes using gene segment 9 and 4 respectively. The use of RT-PCR as a diagnostic tool for group A rotavirus and its comparison with other routine methods such as ELISA and PAGE has been reported. The investigators reported RT-PCR as the most sensitive and specific assay for detection of group A rotaviruses than PAGE and ELISA.

Cultivation of rotavirus in MA 104 cell line   Important progress has been made in the cultivation of the fastidious HRVs directly in tissue culture from clinical specimens by pre-treatment of the specimen with trypsin and incorporation of trypsin into the maintenance medium, use of roller tube cultures of MA 104 cells, and incubation at 37°C. The pre-treatment of the virus with trypsin greatly augmented the ability of the virus to infect cells. Culture is most frequently performed with the monkey kidney epithelial cell line MA 104, although a variety of epithelial cells are permissive for rotavirus growth.

RETROVIRUSES

Human T-Cell Lymphotropic Virus (HTLV) Type 1 and 2

HTLV was first discovered in 1979 by Gallo and colleagues in adult T-cell leukaemia (ATL) with Pautrier's micro abscesses. HTLV-I has been found in association with ATL, HTLV-1 associated myelopathy/tropical spastic paraparesis (HAM/TSP). The viruses are composed of ssRNA with diploid genome, replicates through DNA intermediary and able to integrate into host T-cell genome as a provirus.

Three strains were identified in HTLV-1 viz. Cosmopolitan, Melanesian and Zairian. Cosmopolitan strain has widest distribution worldwide and has three subtypes A, B and C, of which A and B types are found in India.

HTLV-1 is 100 nm in diameter with envelope and spherical core. Genome contain ~9 kb with long terminal repeats (LTR) at 5´ and 3´ ends which code for large overlapping polyprotein processed by virally encoded protease and cellular protease into functional peptides. Region pX code for *Tax* and *Rex* proteins which regulate virus gene expression (figure 4.21). pXORFI and pXORFII are involved in cell cycle regulation. *Gag* protein forms structural proteins of matrix, capsid and nucleocapsid, *Pol* gene encodes several enzymes (protease/polymerase/integrase), *env* (envelope), and a series of accessory proteins that regulate gene expression. The *env* gene encodes surface glycoprotein of 46 kDa (gp46) and the 221 kDa transmembrane glycoprotein (gp21).

Replication cycle of HTLV begins with its attachment to a cell surface receptor followed by uncoating. Studies suggest that preferential infection of CD4 T–helper cells occur via

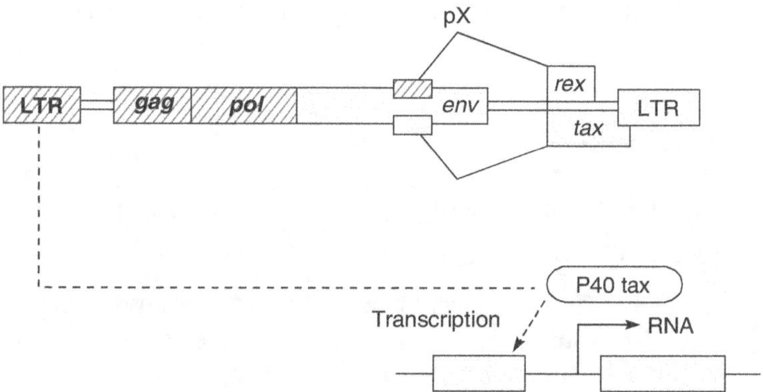

Figure 4.21   Genomic structure of HTLV-I and HTLV-II. The long terminal repeat (LTR) is recognized into three regions— US, R and U3 which house the polyadenylation site; *rev* (response element); and 21-bp enhancer transactivating virus expression. *Gag* (group specific antigen) whose products form the skeleton of virion (matrix, capsid, nucleocapsid, nucleic acid binding protein); *pol*, gene for reverse transcriptase, integrase and protease; *env*, envelop gene; *tax*, transactivating gene; *rex*, viral regulatory gene involves in promoting genomic RNA production.

transcriptional factors. HTLV–II preferentially targets CD8 cells. After virus uptake, viral RNA genome gets reverse transcribed into DNA. This dsDNA is then integrated into host cell genome by the virally encoded integrase, resulting in infection. *Tax* activates transcription of viral genome and *rex* modulates processing of viral pre-mRNA. *Tax* protein plays a major role in viral growth and disease pathogenesis, and is associated with the rapid upregulation of interferon gamma and deregulation of genes for IL-2 and IL-2R. This facilitates extensive cell proliferation.

HTLV-1 epidemiology includes ATL, HAM/TSP and infection via perinatal, postnatal and adults sexually acquire the infection. Peak of infection was observed in 30 to 50 years of age group individuals. Injectable drug users (IDU) constitute majority of HTLV-2 infection. Vertical transmission is the principal mode of childhood infection. Transmission through breast milk of infected mothers is high. Transmission efficiency is high in sexual mode and blood transfusion. Co-factors include elevated virus load heterosexual, ulcerative genital lesion, sharing of needles and paraphernalia.

Clinical manifestations   Three to five percent of the victims develop clinical disease and takes 18 weeks with HAM/TSP to many decades with ATL. ATL is presented with aggressive leukaemia of mature T-cells, generalized lymphadenopathy, visceral involvement, hypercalcemia, cutaneous skin involvement, lytic bone lesion and leukaemia cells. ATLs presents as a non-Hodgkin's T-cell lymphoma with no other clinical factor.

HAM/TSP is associated with demyelination of motor neurons of spinal cord mostly affects adult females. Signs include stiff gait, progressing to increasing spasticity and lower extremity weakness, back pain, urinary incontinence and impotence in men. Sensory symptoms include tingling, pins and needles, burning, and have HTLV-1 antibodies in blood and CSF.

Laboratory diagnosis   Isolation of virus in cell cultures—co-cultivation of patient cells with human peripheral blood mononuclear cells stimulated in vitro with mitogen (e.g. phytohemagglutinin) and IL-2. Antigen capture assay, nucleic acid detection by PCR and antibody detection by ELISA.

Prevention and treatment   Prevention measures include elimination of breast feeding in known infected mothers, use of condoms by discordant couple, screening of blood donors. ATLs are treated with anti-tumour chemotherapy.

Human Immunodeficiency Virus (HIV)

Classification and properties of HIV The family *Retroviridae* is so named for its possession of a reverse transcriptase (Latin, *retro* = backwards). Of the 7 genera it is recognized that only 2 genera cause disease in humans: BLV-HTLV retroviruses which contain HTLV-I and HTLV-II; Lentivirus containing HIV-1 and HIV-2 (Latin, *lentus* = slow) distinguished by the lengthy and insidious onset of clinical signs.

The Spumaviruses (Latin, *spuma* = foam) cause a characteristic foamy appearance in infected primate cell cultures. They are not pathogenic.

HIV, the causative agent of a life-threatening disease, acquired immunodeficiency syndrome (AIDS). Since its first documentation in 1981, the virus has affected almost 50 million individuals by the year 2001 as per the figures released by the joint United Nations Program on HIV/AIDS (UNAIDS) and the World Health Organization (WHO). A major proportion of the infection worldwide is caused by the HIV-1 virus, which was identified in 1983. The type 2 (HIV-2) virus was first detected in West Africa and is significantly prevalent in those regions, Portugal and India with rare cases reported in western countries, Korea and Philippines.

Reacting to the global reports of HIV/AIDS, Indian Council of Medical Research (ICMR) initiated surveillance for HIV infection in India in 1985–86 and the first evidence of HIV infection in sex workers in Chennai, Madurai and Vellore was obtained in 1986–87. Gradually centres were established in the state capitals of India. National AIDS control organization (NACO) set up under the Ministry of Health and Family Welfare mandated to implement initiatives like establishing HIV testing centres, strengthening blood-safety and controlling hospital infection took over surveillance in 1992.

Indian Scenario of HIV India has about 5.1 million people living with HIV next only to South Africa as estimated by National AIDS Control Organization (NACO) by 2003. The national HIV prevalence is between 0.4 and 1.3 per cent. The geographic distribution of the HIV epidemic in India is varied and based on the prevalence of HIV infection in the low and high risk groups, various states have been categorized in high, medium and low prevalence areas. Andhra Pradesh, Tamil Nadu, Maharashtra, Karnataka, Goa and Manipur are classified as high prevalence states with HIV prevalence over 1% even in low-risk populations.

Both HIV-1 and HIV-2 have been shown to exist in India. In western India, HIV-2 and dual infections with HIV-1 and HIV-2 have been reported since early 1990s, and in south India in late 1990s. However, in Kolkata in eastern India, HIV-2 has been only recently detected. Heteroduplex mobility assay (HMA), peptide enzyme immunoassay (PEIA) and DNA sequence analysis have been used for subtype analysis and HIV-1 C has been the most commonly reported subtype in India.

HIV-1   The data from three centres in India show a high prevalence of subtype C. In north India 78.4% of the strains were subtype C. A study undertaken among female sex workers in Kolkata showed 95% of strains to be subtype C. Among these, 68% showed maximum homology to the C3-Indian reference strain (NIH, HMA panel). A study from South India also, showed similar findings with 95% of the 83 strains of subtype C, of which 90.38% were of C3. Table 4.14 summarizes some milestone achievements in area of HIV/AIDS in India.

Table 4.14   HIV/AIDS epidemic in India at a glance.

Year	Comments
1985–86	Indian Council of Medical Research (ICMR) initiated HIV surveillance at specified sites. First report of HIV infection in sex workers in Chennai and first report of AIDS in Mumbai.
1989	HIV infection reported among intravenous drug users (IDUs) in Manipur.
1990–91	National AIDS Control Organization (NACO) was established. Indian National AIDS Control Programme was launched.

(Contd.)

Table 4.14   Contd.

Year	Comments
1992	ICMR established National AIDS Research Institute (NARI) in Pune.
1998–99	Majority of IDUs investigated in Manipur were found to be HIV infected, sexual transmission to partners of HIV infected drug users were documented.
1999	The Supreme Court ruling made HIV testing of all blood bottles mandatory. NACO estimated 3.87 million HIV infections in 1998 in India.
2000–01	Feasibility studies for prevention of parent-to-child transmission of HIV with zidovudine and nevirapine were initiated by NACO.
2001	Anti-retroviral drugs prevailed in the market with reduced cost price.
2002–03	Tripartite agreement between NACO, ICMR and International AIDS vaccine. Initiative signed to facilitate HIV vaccine development and testing in India.
2003	The Central Government announced the policy to provide highly active anti-retroviral therapy (HAART) to those who suffer from AIDS.
2004	NACO implemented phased scale-up programme of anti-retroviral therapy. NACO estimates 5.1 million persons living with HIV in India by October 2003.
2005	First AIDS vaccine trial was initiated in India.

Morphology of HIV   HIV is a spherical enveloped virus, about 90–120 nm in diameter (figure 4.22). It has a unique three layered structure. The innermost is the genome—nucleocapsid complex. It is enclosed within a cone shaped capsid which in turn is surrounded by a matrix protein followed by a host cell membrane derived lipid bilayer envelope that contain glycoprotein spikes of 72 peplomers. The genome is diploid, composed of two identical copies of (+) ssRNA molecules of 9.2 kb each which are associated with reverse transcriptase enzyme. The genome of HIV contain three major genes, viz. *gag* (group specific antigen) gene encodes the core capsid and matrix proteins, the *pol* gene encodes the reverse transcriptase and the *env* gene encodes the virion envelope peplomer protein and transmembrane protein (Table 4.15) (figures 4.23 and 4.24) and antigens of HIV-1 and -2 shown in Table 4.16.

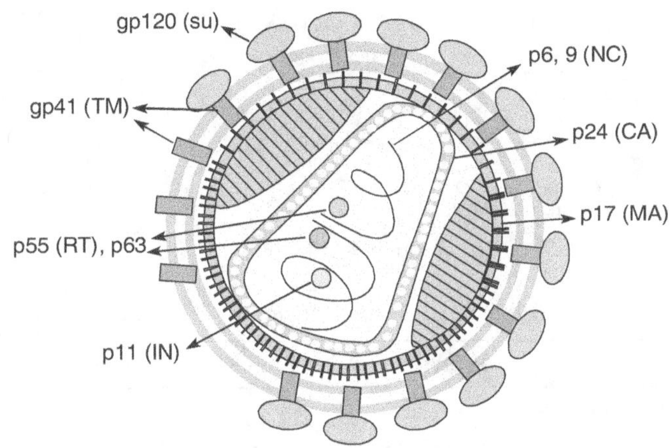

Figure 4.22   HIV Morphology.

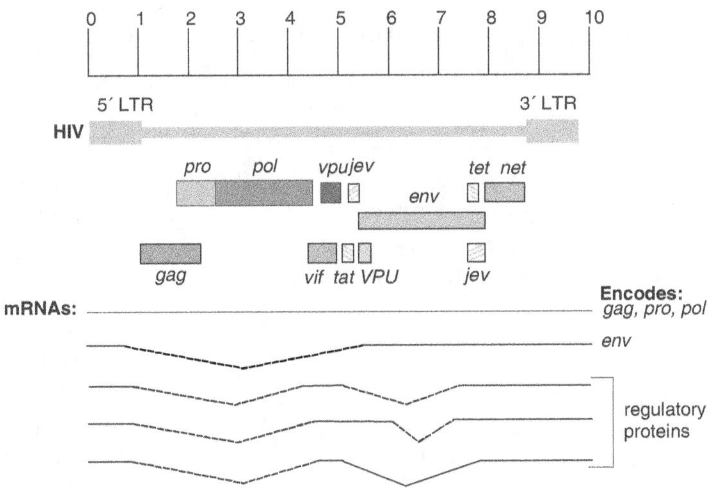

Figure 4.23   HIV genome.

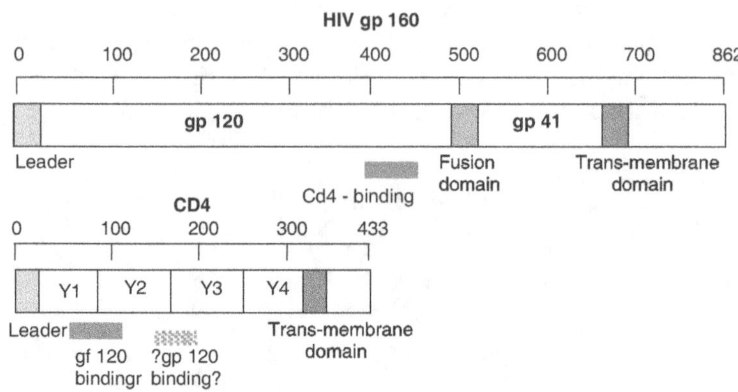

Figure 4.24   HIV *g*P 120.

Table 4.15   *Gene and gene products of HIV.*

Gene	Protein	Size	Function
Structural			
Gag	Matrix (MA)	p17	Structural protein
	Capsid (CA)	p24	
	Nucleocapsid (NC)	p6, p7	
Pol	Protease (PR)	p12	Viral enzyme
	Reverse transcriptase (RT)	p66/p51	
	Integrase (IN)	p32	
env	Surface glycoprotein (S)	gp120	Viral envelope glycoprotein
	Transmembrane protein (TM)	gp41	
Regulatory			
tat	Tat	p14	Transactivation of viral transcription
Rev	Rev	p19	Transport unspliced mRNA to cytoplasm
Vif	Vif	p24	Promote virion and infectivity

Table 4.15   Contd.

Gene	Protein	Size	Function
Nef	Nef	p27	Downregulates Class I MHC and CD4; enhances viral infectivity; facilitate T-cell activation
Vpu (Present only in HIV-1)	Vpu	p16	Promotes release of viral particles; CD4 degradation
Vpr	Vpr	p15	Arrests cell cycle in G2 phase; increase nuclear entry of the preintegration complex in non-dividing cells
Vpx (Present only in HIV-2)	Vpx	p14	Facilitate viral replication and entry of preintegration complex in macrophages

Table 4.16   Major antigens of HIV-1 and HIV-2.

Antigens		HIV-1	HIV-2
A	Envelope antigens/spike antigen	gp120	gp140
	Transmembrane antigen	gp41	gp36
B	Matrix protein antigen	p17	p16
C	Core antigen	p24	p26

Molecular characterization   Based on the heterogeneity of the HIV RNA, three groups are recognized in HIV-1: M (major/main), N (non-M, non-O/new) and O (outlier). The *env* proteins of group M and O can show a variation as much as 30–50%. The N subtype appears to be phylogenetically equidistant from M and O.

The M group is the most prevalent among the three groups. It has nine subtypes (A–D, F–H, J, K) all of which have originated from Central Africa. Within subtypes A and F there are separate sub-clusters designated A1, A2 and F1, F2 respectively. Group M strains of HIV-1 is prevalent all over the continent, while N and O are restricted to Central Africa. Subtype B is the most prevalent form in western and central Europe, Americas and Australia. Subtype C is most prevalent in Indian subcontinent. The strain analysis found 68% of the circulating subtype to be C3.

Modes of transmission of HIV   There are three modes of transmission: sexual, parenteral, and perinatal. Of these, sexual mode is the most important and most common one reported with ~80%, and 10% each of parenteral and perinatal mode. Sexual behaviour as reflected by more number of sexual partners, sex with sex workers, homosexuals (men having sex with men—MSM), being in sex work and receptive anal sex, has been reported to be strongly associated with HIV infection. Risky sexual behaviour of mobile and travelling populations has been linked to HIV infection.

Parenteral transmission may occur through blood transfusion, and blood products. This occurs when blood is collected during the "window" period of infection, that is the interval between the time of exposure to the virus and development of antibodies.. Therefore screening of blood is of high priority in controlling the transmission of the virus. Infection can be transmitted by blood products like plasma, serum and cells from HIV positive individuals, also from organ donations from these AIDS (+) cases.

HIV is also widely transmitted among injecting drug users (IDUs). The problem of HIV among IDUs in Manipur increased rapidly from 2–3% in 1989 to >50% in 1991 and has been estimated to be above 75% at present, shows the possibility of HIV transmission through sharing of needles routinely share injecting equipment.

Perinatal transmission may occur from parent who is HIV +ve to the child during pregnancy. There is a potential for perinatal HIV transmission resulting in paediatric AIDS epidemic. Studies from India documented mother-to-infant HIV transmission rates between 36% and 48%. It has been reported that 63% of the multi-transposed thalassaemic HIV infected children became symptomatic in infancy with a 9% fatality rate within 14 months of age.

HIV and sexually transmitted diseases (STDs)   Among the various factors associated with the sexual transmission of HIV, STDs seem to contribute significantly. WHO estimated that there were 333 million new cases of the four curable STDs, worldwide in 1995 among adults (15–49 yr). In developing countries, both prevalence and incidence of STDs are very high.

The AIDS Prevention and Control (APAC) project jointly undertaken by Voluntary Health Services (VHS) and US Agency for International Development (USAID) in a community of urban, rural adult population of Tamil Nadu revealed that the HIV prevalence in the study community was 1.8% and that of any STD was 15.8% with the classical STDs being 9.7%. Genital discharge was the most common STD syndrome (41.5%) in women and trichomoniasis (5.1%), chlamydia (3.9%), and gonorrhoea (3.7%) were the most common etiologies. The age group of 30–39 years was at the highest risk for any STD.

Clinical profile of HIV in India   The prolonged course of HIV infection is marked by a decrease in the CD4+ count and persistent viral replication, resulting in immunologic decline and death from opportunistic infections and neoplasms. Acute infection is characterized by a rapid rise in plasma viraemia with a concomitant drop in CD4 count within 3–6 weeks of exposure. Associate symptoms include fever, sore throat, skin rash, lymphadenopathy, splenomegaly, myalgia, arthritis, and less often, meningitis. This is followed by a clinically latent period with low level viral replication and a gradual fall in CD4 count where patient can remain asymptomatic for several months to years. Mean duration of survival after diagnosis with HIV in India is 7 yr 8 months. Median time for progression from HIV infection to AIDS was 7.9 yr in one study of patients from Mumbai. With CD4 counts <200 cells/μl, patients are at high risk for developing opportunistic infections (OIs) like tuberculosis (TB), Pneumocystis carinii pneumonia (PCP), toxoplasmosis, and cryptococcal meningitis. Specific AIDS defining illness, CD4 counts, and HIV RNA levels predict survival of patients with HIV infection. Disease progression correlates with clinical features such as chronic fever, persistent cough for more than 1 month, chronic diarrhoea, oral candidiasis, severe chronic HSV infection, >10% loss of body weight within 1 month, and incident TB. Patients with CD4 count <200 cells/μl are 19 times more likely to die than those with CD4 count >350 cells/μl.

Pulmonary manifestations of HIV   HIV–TB co-infection is a serious problem worldwide, but is of a major concern in India which ranges from 2.8% to 9.4% among HIV patients. The risk of developing TB after an infectious contact is 5–10% per year among HIV infected individuals compared to 5–10% during the lifetime of HIV-negative individuals. TB is more frequent in patients whose CD4 count is <300 cells/μl.

Pneumocystis jerovecii causes severe pneumonia in patients with AIDS. Occurrence of Pneumocystis jerovecii pneumonia PCP establishes the diagnosis of AIDS and it is the most common AIDS defining illness in the developing world.

An opportunistic infection of bacterial pneumonia was reported in 1.8% of Southern Indian HIV-positive patients. The common causes are encapsulated bacteria, *Streptococcus pneumoniae* and *Haemophilus influenzae*.

Oral lesions in HIV   Oral lesions are important in early diagnosis and to monitor the progression of disease.

Oral candidiasis   It is the most common HIV-associated condition occurring in up to 70% of cases. The median CD4 counts of patients with candida ranged between 107 and 189 cells/μl with a PPV of 75%.

Periodontal disease   Three characteristic presentation are seen— necrotizing periodontal disease, linear gingival erythema (LGE), and exacerbated attachment loss. LGE is described by rapid loss of bone and soft tissue in clean mouths where there is little plaque or calculus to account for the gingivitis.

Oral hairy leukoplakia (OHL)   Prevalence of OHL ranges from 0 to 26%, median CD4 count was <129 cells/μl.

Oral ulcers   This is caused primarily by HSV. HSV lesions are the third most common mucocutaneous lesion in HIV infected individuals after candida and dermatophytosis. Median CD4 count = 219 cells/μl with 5.7% of HIV positive cases having acute ulcers.

Oral pigmentation   Patchy brown to brownish-black asymmetrical lesions usually >1 cm, which are distinctive from racial oral pigmentation, have been reported in up to 23% of HIV positive cases.

Dermatological conditions associated with HIV   Cutaneous manifestations occur in 90% of HIV infected individuals and can be classified into four groups.

Infectious   Herpes zoster, lesions in genital and anal areas, human papilloma virus (HPV) induced oral, genital and anal warts, molluscum contagiosum induced pink papules, staphylococcal skin infection (reported in 1.3% HIV positive patients; mean CD4 count = 410 cells/μl).

Autoimmune   Papular pruritic eruption (PPE) is a unique dermatosis associated with advanced HIV infection, characterized by sterile papules, nodules, or pustules with hyperpigmented, utricarial appearance, and pruritis. Alopecia, vitiligo, psoriasis, eosinophilic folliculitis are all examples of autoimmune conditions in HIV positive patients.

Drug-induced   Toxicides resulting from HIV therapy also induces dermatologic conditions from generalized morbilliform exanthema to severe Steven Johnson's Syndrome (SJS) in spite of safe, tolerable and efficacious generic HAART regimens.

HIV related Cutaneous malignancies   This includes squamous cell carcinoma, basal cell carcinoma, and Kaposi's sarcoma.

Neurological manifestations of HIV   Neurological manifestations are seen in 20% of outpatients in HIV clinics and are categorized into opportunistic infections, malignancy, AIDS related dementia, vasculitis/stroke and psychiatric illnesses.

Opportunistic infections in CNS Cryptococcal meningitis (2–4.7%), diagnosis requires demonstration of organism in CSF, and serum cryptococcal antigen treatment involves intravenous administration of amphotericin B for 2 weeks; toxoplasmosis another common OI of the CNS, others include herpes encephalitis, fulminant pyogenic meningitis, meningococcal meningitis, acanthamoeba infection, aspergillus infection, rhizopus infection and neurosyphilis.

Malignancy Non-Hodgkin's lymphoma (NHL) is the second most common HIV-related malignancy after Kaposi's sarcoma occurring in 2–5% of AIDS cases.

AIDS dementia complex (ADC) Reports about ADC in India are minimal. One study in Jaipur of 30 AIDS patient, 4 of whom were diagnosed with ADC. ADC is important to recognize and manage because it impacts the quality of life of patients and impinges on their ability to function properly. HAART remains the only option for management of ADC.

Vasculitis/stroke HIV itself can cause vascular endothelial damage, predisposing patients with advanced disease to stroke. An autopsy study of AIDS patients revealed 15% of cases with infarcts/haemorrhages.

Psychiatric illnesses HIV/AIDS is confounded by psychiatric illnesses, both pre-existing, and ones that develop after learning of diagnosis. Pre-existing illnesses such as drug and alcohol dependence predispose patients to behaviour that puts them at risk for acquiring HIV infection. Presence of pain, concurrent alcohol abuse, poor family relations, and presence of AIDS in the spouse were significant factors associated with depression, anxiety, and suicidal ideation. Supportive counselling accompanying clinical care is crucial for these patients.

Gastrointestinal manifestations of HIV The gastrointestinal manifestations are used below:

Esophagitis Cause dysphagia or odynophagia. This is very common among patients with advanced HIV disease.

Diarrhoeal diseases   Chronic diarrhoea is a major problem in HIV infected persons seen in up to 76% cases. *Isospora belli* and *Cryptosporidium parvum* were the two most common causes of chronic diarrhoeal disease, followed by *Blastocystis hominis, Strongyloides stercoralis, Entamoeba histolytica, Giardia lamblia,* enteropathogenic *Escherichia coli, Enterocytozoon bieneusi* and *Campylobacter jejuni.*

Abdominal mass lesions   Caused by abdominal tuberculosis and abdominal lymphoma found in patients with diarrhoea, pain, obstruction, bleeding, or perforation.

Hepatitis B and C   Concurrent infection with HIV and HBV/ HCV is of great concern worldwide. In India, rates of co-infection with HIV and HBV are reported between 6% and 33%.

Ocular manifestations of HIV   The most common ophthalmic opportunistic infection in India is CMV retinitis, which almost always occurs in patients with CD4 counts <50 cells/μl, followed by non-infectious retinopathy (HIV retinopathy) reported in 13–15% of HIV patients. Less common ocular infections are toxoplasma gondii, VZV, and pneumocystis carinii.

HIV-associated malignancies   Patients with AIDS are at increased risk for developing NHL, followed by Hodgkin's disease and plasmacytoma. In addition to lymphomas, Kaposi's sarcoma, squamous cell carcinoma, and cervical cancer are malignancies of concern in HIV-infected individuals.

Laboratory diagnosis   Over the past 20 years, HIV diagnostics have been essential in detecting and monitoring infection, and continue to play a major role in saving lives throughout the world. As technology evolved, screening, confirmatory, and HIV monitoring assays have been improved and offer better alternatives to address blood screening, surveillance, diagnosis, and patient management. Molecular methods are critical in detecting early infection and for managing patients on anti-retroviral therapy. Molecular methods such as polymerase chain reaction (PCR), nucleic acid sequence based amplification (NASBA), and branched DNA (bDNA) technology offer increased sensitivity and genotyping to detect mutations associated with drug resistance. Determination of CD4 lymphocyte counts are important to determine the commencement time of anti-retroviral therapy.

Markers of HIV diagnosis   After HIV infection, the sequence of blood markers to identify infection, in their chronological order of appearance, is as follows: viral RNA, p24 antigen, and antibody to HIV antigens. Within 2 weeks after infection, viraemia occurs as measured by viral RNA. Viral RNA levels are an indication of viral replication, usually reach close to 1 million copies of RNA/ml in couple of months and gradually decrease to a fairly constant level known as *set point*. The time interval before antibody appears, known as the serological *window period*, is characterized by seronegativity, detectable viraemia (as measured by RNA or p24 antigen), and variable CD4 lymphocyte levels. Detection of specific antibody to HIV signals the end of window period and labelled as seropositive. Antibody usually appears 3–4 weeks after infection. When CD4 levels fall <200 cells/ml of blood, severe immunodeficiency occurs, the person is labelled as having AIDS.

ELISA for antibody detection include indirect ELISAs and the 3rd generation antigen sandwich ELISAs (advantage is that all classes of antibody, including IgM will be detected), thereby increases the sensitivity of early detection. More recently, 4th generation ELISA tests have been developed that detect both antibody and p24 antigen simultaneously. ELISA is used to detect p24 antigen. It detects the viral capsid (core) p24 protein in blood which is detected earlier than HIV antibody during acute infection. The p24 antigen appears 16 days after infection as a result of initial burst of the virus replication. It is based on antibody sandwich ELISA where specific monoclonal antibodies to HIV p24 are attached to the solid phase and "capture" viral antigen in the sample. An antibody detector (a biotinylated anti-p24 antibody) is added, followed by conjugate addition (streptavidin-horseradish peroxidase). Subsequent substrate addition results in colour production in positive cases.

Confirmatory assays   They are designed to offer a greater specificity than screening assays, and therefore are the methods of choice to verify that reactive results from screening assays represent HIV infection. HIV western blots, line immunoassays (LIA) and indirect immunofluorescence assay (IFA) are the most common serologic confirmatory assays. In western blots, the HIV-1 viral antigens are separated as gp160, gp120, p66, p55, p51, gp41, p31, p24, p17 and p15, and blotted onto nitrocellulose membrane for development. LIA in principle is similar to western blot. The difference is that in LIAs, artificial HIV antigens are "painted" on the strips rather than being electrophoresed from viral lysates.

Nucleic acid tests for HIV include: RT-PCR reverse transcription polymerase chain reaction), nucleic acid based amplification (NASBA) or transcription mediated amplification (TMA), and branched chain DNA (bDNA).

AIDS vaccine development   The main scientific obstacles to the development of an AIDS vaccine are summarized below.

- Antigenic diversity and hypervariability of the virus.
- Transmission of disease by mucosal route.
- Transmission of the virus by infected cells.
- Resistance of wild type virus to seroneutralization.
- Integration of the virus genome into the host cell chromosomes.
- Latency of the virus in resting memory T-cells.
- Rapid emergence of viral escape mutants in the host.
- Downregulation of major histocompatibility (MHC) class I antigens.

The high error rate of the viral reverse transcriptase leads to continuous mutations in the HIV genome. A typical immune response to HIV infection involves the development of both neutralizing antibodies and cell mediated immune responses especially CTLs against CCR5 primary HIV isolates.

Over the past one decade, several developing countries have joined the international efforts of AIDS vaccine development shown in Table 4.17.

Table 4.17 Clinical trials of AIDS vaccines in developing countries.

Year	Country	Phase	Product
1993–94	China, Thailand , Brazil	II	V3 branched peptide
1995–99	Thailand	I/II	rgp 120 B (MN)/(SF2)
		II/III	rgp 120 BE (SF2/CM235)/(MN/A244)
1999	Uganda	II	ALVAC vCP205
2000–03	Thailand	I/II/III	ALVAC vCP1521 + rgp 120 BE
2001–02	Kenya	I	DNA-HIVA, MVA-HIVA
	Haiti, Trinidad, Brazil, Peru	II	ALVAC vCP1452 + rgp 120 BB
2003	Uganda, Kenya, South Africa	II	MVA-HIVA, DNA-HIVA
	Botswana	I	DNA-multiepitope
	Brazil, Peru, Thailand, Haiti	I/II	MRKAd-5 *gag* B
2004	USA, Canada, Peru, Australia	IIb	MRKAd-5 *gag/pol/nef* B

Ad-5—Human adenovirus type 5 recombinant expressing HIV-1 proteins.

AIDS vaccine approaches

- *Recombinant envelope subunits*—Recombinant soluble HIV and SIV glycoproteins have been used as subunit vaccines.

- *Synthetic peptides*—Synthetic peptide immunogens concentrated on the V3 loop of gp120.

- *Live recombinant vectors*—Live attenuated viral or bacterial strains used as a vector to carry HIV genes encoding the antigens of interest shown below.

    - Vaccinia + recombinant envelope subunit

- Canary pox + recombinant envelope subunit
- Salmonella + recombinant envelope subunit
- DNA + pox virus (canarypox, fowlpox, MVA (modified vaccinia Ankara), a highly attenuated, host range-restricted strain of vaccinia virus
- DNA + adenovirus
- AAV (adeno associated virus is a naturally occurring virus that depends on a helper virus for replication) + homologous or heterologous AAV or MVA

In India one centre is located at the National AIDS Research Institute (NARI), Pune, working towards conducting the phase I clinical trial with the AAV-based AIDS vaccine in 2005. The second AIDS vaccine clinical evaluation centre is being set up at the Tuberculosis Research Centre (TRC), Chennai. A phase I trial with the MVA vaccine is expected to begin in 2005.

MYXOVIRUSES

ORTHOMYXOVIRIDAE

Members of *Orthomyxoviridae* are spherical or filamentous with single-stranded (ss) and segmented RNA genome. They are so called because of their affinity to mucins. This family includes three genera influenza A, B and C. They infect humans, birds, pigs and horses.

INFLUENZA VIRUS

Morphology

Influenzavirus A and B are morphologically similar but C has certain differences in certain respects in particular having only a single glycoprotein spike. The virions are spherical, 80–120 nm in diameter (figure 4.25), but larger pleomorphic and filamentous forms 1000 nm or more in length may be abundant. The nucleocapsid has helical symmetry (figure 4. 26) with a core of 8 segmented ssRNA (–)-sense of ~9kb (about 5000 kDa). RNA genome is associated with an RNA polymerase (PA, PB1, PB2) proteins to form a ribonucleoprotein (RNP) arranged in a helix. The nucleocapsid is surrounded by a matrix protein (M1) layer, which in turn is surrounded by host plasma membrane derived lipid bilayer envelope (figure 4.27). The M2 protein projects through the envelope to form ion channels which allow pH changes in the endosome.

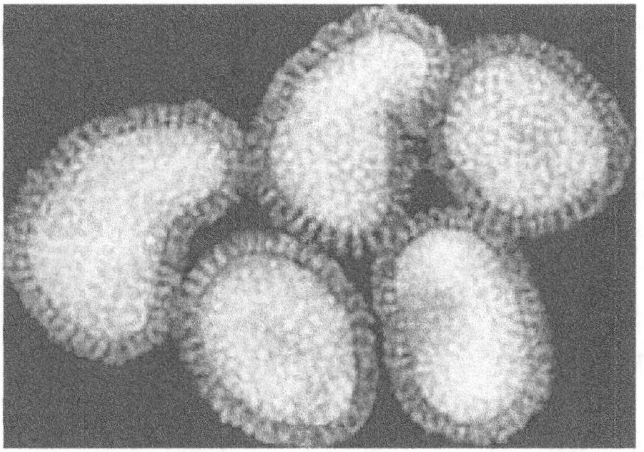

Figure 4.25   Influenza virus.

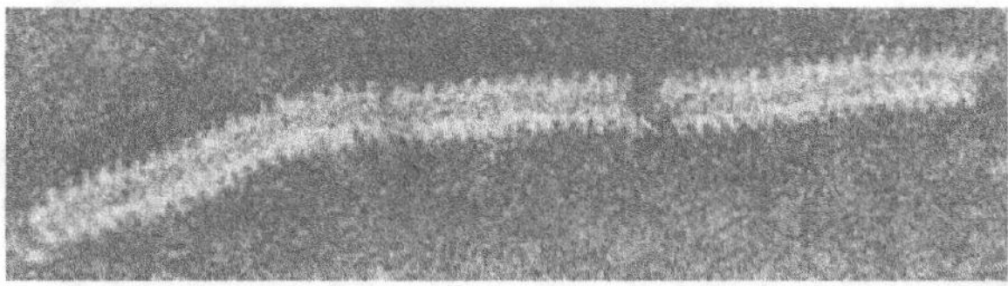

Figure 4.26   Helical RNP.

The envelope of influenza virus A and B has two types of glycoprotein projections or spikes, the haemagglutinin (H) and the neuraminidase (N) proteins. The 'H' peplomers exist as 500 tapered projections, and are 10 nm in length with their broader ends (5 nm) outside and the narrower ends inserted into lipid membrane. The 'N' peplomers are 100 mushroom shaped projections per virion measuring 9 nm in length. The virus particles bind to host cell through sialic acid receptor by 'H' peplomers. This H peplomer causes agglutination of RBCs an important reaction made use of to type the virus in cell cultures. Haemagglutination is followed after some time by detachment of the virus from the cell surface, this is caused by enzyme neuraminidase present in the 'N' peplomers.

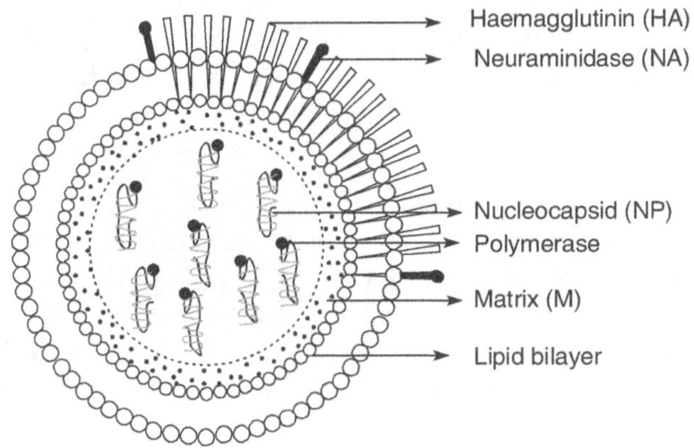

Figure 4.27   Schematic presentation of an influenza virus particle.

Influenzavirus A is divided into subtypes, based on their differences in H and N proteins. Fourteen subtypes of H protein (H1–H14) and 9 of N proteins (N1–N9) are known in birds. Some of which are found in various combinations in mammals including humans. Genome of influenza virus C contains only 7 segmented RNA and has no 'N' activity.

Antigenic Structure

Influenza virus, antigenic structure include haemagglutinin (H) glycoprotein composed of two polypeptides HA1 and HA2 strain specific, neuraminidase (N) glycoprotein enzyme which destroys cell receptors by hydrolytic cleavage and strain specific like H, matrix (M) protein which is genus specific, and ribonucleoprotein (RNP) that is genus specific.

Replication

After attachment to the specific receptors on the cell membrane, virions are taken into the cell by endocytosis and then transported to vacuoles (endosomes), where the acid pH induces a change in the configuration of the HA. This rearrangement of the HA brings a special set of catalytic amino acids, the 'fusion sequence', in contact with the lipid of the vacuole wall. Fusion of the viral and vacuole lipids triggers release of the virus ssRNA, which is transported immediately to the host-cell nucleus. Simultaneously protons pass along the M2 ion channel to the interior of the virion and cause M1 protein to be released from the RNP complex. Thus the RNP : RNA complex can enter the cell nucleus for transcription and replication.

The new influenza virions are assembled at the host cell surface membrane and released by a process of budding in which both HA and NA are involved. Viral NA has the important function of cleaving sialic acid from viral and cellular glycoproteins, thus preventing virus aggregation and allowing individual virions to be released from the cell.

Antigenic Variation

Influenza virus is capable of undergoing antigenic variation due to dynamic changes in the antigenicity of H and N proteins. Depending on the degree of antigenic changes occurring in H and N, two distinct forms are known.

Antigenic drift   The gradual sequential change in the antigenic structure occurring regularly at frequent intervals is known as antigenic drift. It results from mutations in H or N genes. The new antigens acquired though different from the previous are yet related to them so that they can still react with antisera to the predecessor virus strain.

Antigenic shift   This is an abrupt, drastic, discontinuous variation in the antigenic structure resulting in a new virus strain unrelated antigenically to predecessor virus strain. It may involve both H and N genes. This is a result of genetic recombination in doubly infected cells.

Pandemic influenza   The great pandemic of Spanish influenza in 1918 was especially terrible in its effects in Europe; worldwide, it killed about 30 million people (Table 4.18). An influenza pandemic that occurred nearly 30 years later in 1957, when a strain differing completely in both HA and NA appeared in China, was less virulent. In 1968, there was another pandemic, again originating in the far-east; the virus, first isolated in Hong Kong, had now undergone a partial shift that affected only the HA and not the NA.

Properties

It can be inactivated by heat (56°C for 30 min.), acid pH 3.0, ether, phenol, formaldehyde, detergents, soaps, iodine, etc. It remains viable at 0–4°C for a week and can be preserved at –70°C for long periods and indefinitely by freeze-drying. It can be isolated in the amniotic cavity of 13-day old chick embryos incubated at 33–35°C for 2–3 days. Isolation of the virus can also be done on primary monkey kidney or human embryonic kidney cells. Madin–Darby canine kidney (MDCK) cells, a continuous cell line, can be used to isolate influenza viruses, provided trypsin (2 μg/ml) is present in the medium to cleave the haemagglutinin of progeny virions for better spread.

Table 4.18   Strains of influenza virus that caused pandemics.

Virus strain	Pandemic year	Drift/shift
H1N1	1918	New pandemic shift virus HA/NA drift
H2N2	1957	New pandemic shift virus HA/NA drift
H3N2	1968	New pandemic shift virus
H1N1	1976/77 2002	HA/NA drift

Pathogenesis

The route of entry of the virus is the respiratory tract. The viral neuraminidase facilitates attachment by hydrolyzing the mucous film lining the respiratory tract exposing the cell receptors. Incubation period is short varying from 1–4 days. The illness is characterized by a sudden onset of systemic symptoms such as chills, fever, sore throat, non-productive cough, myalgia, headache and malaise. The uncomplicated illness usually lasts for 3–7 days.

Complications Young children may develop croup, pneumonia or middle ear infections. Elderly may develop secondary bacterial pneumonia due to *Staphylococcus aureus*, *Streptococcus pneumoniae*, and *Haemophilus influenzae*, and Reye's syndrome characterized by severe encephalomyelitis accompanied by liver degeneration.

Laboratory Diagnosis

1. Demonstration of viral antigens in smears of nasopharyngeal secretions, centrifuged deposit of throat garglings by FAT (fluorescent antibody test using monoclonal antibodies).

2. Isolation of virus in amniotic cavity of chick embryo, MDCK (Madin-Darby Canine Kidney) cells, primary monkey kidney or human embryonic kidney cells with specimens collected during first 2–3 days of illness. Throat garglings are collected in sterile buffered solution and processed for the assay.

3. Serological diagnosis of influenza is based upon demonstrating a 4-fold or higher increase in antibody titre between acute phase and convalescent phase sera. Complement fixation, haemagglutination inhibition and neutralization tests are done for this purpose.

Prophylaxis

Vaccination is the main preventive measure for influenza, but difficulty in immunoprophylaxis attribute to frequent antigenic variation. Both inactivated and live vaccines are available. Inactivated vaccines are prepared by allantoic cavity grown virus inactivated with formalin or b-propiolactone, then purified by ultracentrifugation and extracted with ether and detergent to reduce the side-effects of the whole-virus vaccine ('split' vaccine). The resulting polyvalent vaccine is used every autumn. This vaccine can be further purified to contain only H and N proteins, which is known as 'subunit' vaccine (Table 4.19).

Live attenuated vaccines include temperature sensitive mutants of influenza virus that grow at lower temperature of the nasopharyngeal mucosa (32–34°C) but not in lungs (37°C). Live vaccine is administered by aerosol spray or intranasally. It stimulates local IgA response.

Table 4.19   Inactivated influenza vaccines.

Type	Remarks	Suitable for children
Whole virus	Good immunogen but may cause local and general reactions	No
'Split vaccine'	Whole virus extracted with ether	Yes
'Subunit' or 'Surface antigen'	Purified HA and NA extracted with detergent; less liable to cause reactions	Yes

Chemoprophylaxis   Rimantadine, a methylated derivative, and amantadine, a primary amine may be given orally to unimmunized high risk people during a major epidemic of influenza virus A.

PARAMYXOVIRIDAE

Members of the family *Paramyxoviridae* are enveloped, negative-stranded ssRNA viruses, 150–200 nm in diameter, with a

Table 4.20 Members of Paramyxoviridae family.

Subfamily	Genus	Human species
Paramyxovirinae	*Paramyxovirus*	Human parainfluenza virus types 1 and 3
	Rubulavirus	Human parainfluenza virus types 2, 4a, 4b, mumps virus
	Morbillivirus	Measles virus
Pneumovirinae	*Pneumovirus*	Human respiratory syncytial virus

nucleocapsid of helical symmetry. Two subfamilies are recognized: the Paramyxovirinae, containing the genera *Paramyxovirus*, *Rubulavirus*, and *Morbillivirus*, and the *Pneumovirinae*, with one genus, *Pneumovirus* (Table 4.20). The morbilliviruses differ from those in the other genera in not possessing a neuraminidase.

Morphology and Structural Proteins

Because of the fragility of the lipoprotein envelope, the viruses often appear distorted or disrupted in the electron microscope, with the nucleoprotein spewing out from inside the virion (figure 4.28). The structural proteins of the capsid include the HN (haemagglutinin-neuraminidase) and F (fusion) glycoproteins which form the surface spikes, and the internally situated M (matrix) protein. Three other proteins, together with the RNA, form the RNP core of the virion and the RNA transcriptase possessed by all negative-stranded viruses (figure 4.29). The F protein, formed by proteolytic cleavage of a larger precursor polypeptide, is important, since it mediates the fusion of infected cells which then form the syncytia so characteristic of infections with this group of viruses.

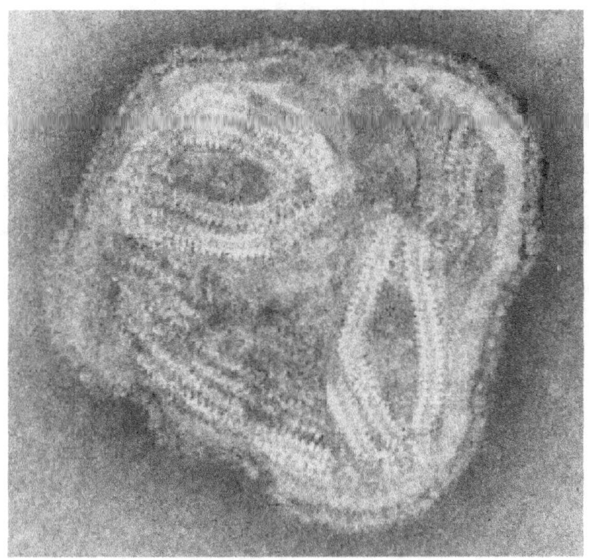

Figure 4.28   Paramyxovirus.

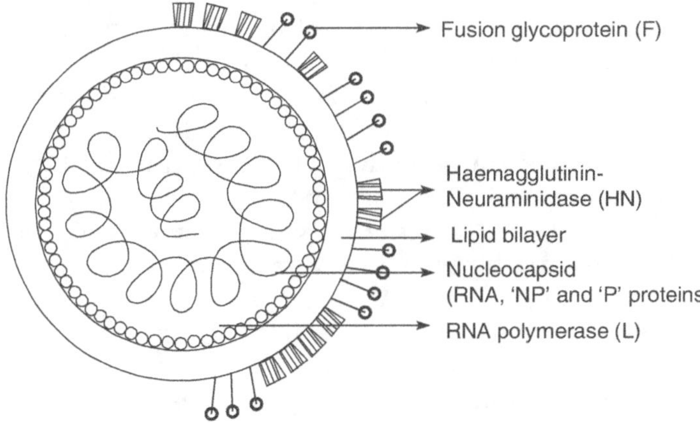

Figure 4.29   Schematic diagram of the probable arrangement of the structural components of a paramyxovirus.

Genome

The genome is an ssRNA molecule of negative polarity of some 15,000 nucleotides, containing the 6–10 genes coding for the 6–10 known virus specific proteins (figure 4.30).

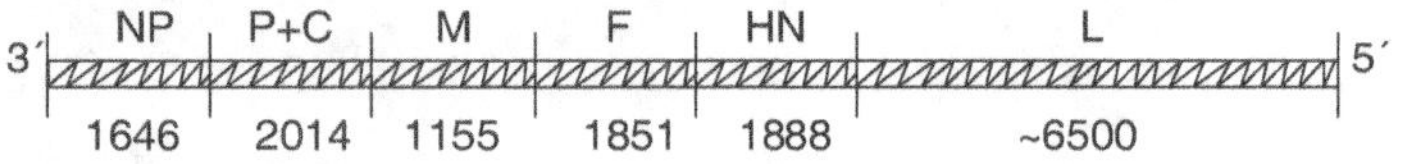

Figure 4.30   Schematic presentation of the genome of parainfluenza virus type 3. NP - Nucleoprotein; P+C - Small proteins of polymerase complex; M - Matrix; F - Fusion protein; HN - Haemagglutinin-neuraminidase; L - Large protein of the polymerase complex.

Replication

The strategy of replication is that of a typical negative-stranded RNA virus. These viruses replicate entirely within the cytoplasm of the cell. Viruses dock to the cell using a glycoprotein or glycolipid cell receptor and its own G, H, or HN spike protein. The virion enters a cell by mediating fusion of its lipid envelope with the external plasma membrane of the cell and this vital event of "catalysed fusion from without" is mediated by F protein. The genome is released immediately into the cytoplasm. The genome is transcribed by viral RNA transcriptase to give 6–10 subgenomic, generally monocistronic mRNAs. Shortly, afterwards gene transcription switches into gene replication, when full-length positive RNA strands are synthesized which act as templates for RNA replication. The newly synthesized (–) RNA strands interact with N protein and the virion transcriptase, and these nucleocapsid structures then combine with viral M protein. The viral spikes already inserted into the plasma membrane of the cell interact with the M protein. The viruses are released from the cell by budding from the plasma membrane. Because of the tendency of the infected cells to fuse, they may, under certain conditions, release hardly any new viruses into the extracellular environment; instead, the viruses may spread from cell to cell, behaving as typical 'creepers'.

Clinical Manifestation

The signal illness for parainfluenza virus (PIV) is croup, or laryngotracheobronchitis. Croup is manifested by fever, hoarseness, and a barking cough in a child usually between 6 and 8 months of age. Progression of narrowing of subglottic area of the trachea may occur. PIV type 1 is the major cause of croup. PIV type 2 is also associated with croup. PIV type 3 is a cause of sporadic croup and is also associated with bronchiolitis and pneumonia in young infants. Complications due to PIV include otitis media, bacterial pneumonia.

Pathogenesis

PIVs are highly transmissible and are acquired by aerosolization of respiratory secretion. Incubation period varies from 2–6 days. All PIVs produce upper respiratory tract infections, in infants and young children they may cause lower respiratory tract infection with pneumonia. In older children (6 months to 5 years) they cause croup.

Laboratory Diagnosis

The clinical specimens for diagnosis include nasal wash or aspirate, bronchoalveolar lavage (BAL). Antigen detection is done by ELISA , isolation in cell cultures using primary monkey kidney cells and confirmed by haemadsorption inhibition test. Immunofluorescence and hemagglutination inhibition tests are also employed. Detection of PIV-RNA is done by RT-PCR.

Measles

"Measles" derived from an Anglo–Saxon word, *measles* and so has obviously been with us for many centuries. Its Latin name, *morbilli,* is a diminutive of morbus, a disease, and thus signifies a minor illness. In tropical areas it is a killer of children on a large scale.

Measles virus (MV) is a member of the genus *Morbillivirus* within the family *Paramyxoviridae*, differ from other paramyxoviruses in that they do not have any detectable neuraminidase activity. In addition, the formation of intranuclear inclusion bodies is a distinctive feature of their cytopathology. There is only one serotype of MV and recovery from natural measles confers life long immunity to the disease.

Composition of the virus   MVs are pleomorphic, enveloped particles with a helical nucleocapsid, morphologically indistinguishable from virions of other paramyxoviruses. Diameter varies from 100 to 250 nm, comprises of bilayered envelope with surface projections (peplomers) with a length of 9–15 nm. It is composed of two different transmembrane glycoproteins, the 'H' and 'F' proteins. H peplomers appear conical, whereas F peplomers resemble dumbbells with ends of unequal size. Matrix (M) protein on the inner side interacts with cytoplasmic tails of H and F. Helical nucleocapsid is packed in the form of a symmetrical 2500 copies of nucleoprotein (N) bound to genomic RNA, together with large polymerase protein (L). Nucleocapsid is of 17–18 nm diameter with a central hole 5 nm in diameter and a pitch of ~5 nm. Genome is a linear ssRNA of (+)-sense with an estimated molecular weight of 4.5×10^6 Daltons. Six measles virus structural proteins has been identified, three proteins N (nucleoprotein), P (phosphoprotein) and L (large protein) are complexed with viral RNA to form the nucleocapsid, and three proteins M (matrix protein), H (hemagglutinin), and F (fusion protein) participate in the formation of virus envelope. Two non-structural proteins C and V identified in cytoplasm of infected cells are involved in regulation of viral RNA synthesis and genome replication respectively.

Transmission   Principal mode is via large droplets of infected respiratory tract secretions inhaled as a consequence of face-to-face exposure to individuals during the catarrhal stage of the disease. Virus persists during incubation period, prodromal and for first 2–3 days of the rash. Viraemia is present during this time; virus is also shed in urine during prodrome and for 6 or more days after onset of rash.

Incubation period is generally 8–12 days, typically about 10 days in children, may be shorter in infants and as long as 3 weeks in adults. First 2–4 days after infection, MV proliferate locally in the respiratory mucosa and spreads to draining lymph nodes where replication occurs and expresses leucocyte function associated antigen-1 (LFA-1). The virus then enters bloodstream causing primary viraemia, then disseminated to reticuloendothelial system. Secondary viraemia occurs 5–7 days after infection and virus spread into tissues in the body, and then prodromal symptoms and signs appear in about 8–12 days after infection.

Classic measles   The prodromal phase of measles begins after an incubation period of 8–12 days with fever, malaise, and anorexia followed in a few hours by coryza, conjunctivitis, sneezing and cough.

The catarrhal symptoms increase in intensity, coryza is intense, with a profuse mucopurulent nasal discharge, there is palpebral conjunctivitis with lacrimation. The cough is severe with brassy, barking quality. 2–3 hours before onset of rash, Koplik spots appear on the buccal mucosa. Koplik described them as small (1–3 mm) irregular bright red spot with a minute bluish-white speck at its centre.

Clinical features   A prodromal stage lasting 2–3 days, with running eyes and nose, cough, and moderate fever. Examination of the buccal mucosa adjacent to the molar teeth may reveal Koplik spots, which resemble grains of salt just beneath the mucosa. They may be many or few in number.

The rash first appears on the face and spreads to the trunk and limbs within the next 2 days. It is a dull-red, blotchy, maculopapular eruption, usually characteristic of its name "morbilliform". There is evidence of bronchitis and pneumonitis, with cough and 'crackles' in the chest. Complications include acute post infectious encephalomyelitis, subacute sclerosing panencephalitis (SSPE) and subacute measles encephalitis (MIBE). Diarrhoea has also been reported in few cases.

Laboratory diagnosis   Isolated from throat washings, blood, urinary sediment, on monkey kidney, primary human embryonic kidney cells. The appearance of multinucleated giant cells containing numerous eosinophilic inclusions in cytoplasm and nuclei seen in measles virus infection. Serology include detection specific IgM antibodies by ELISA. Detection of MV RNA by RT-PCR using primers targeted to N, M or F genes.

Prophylaxis   Children of the age of 15–18 months of age are given MMR vaccine, followed by a booster at the age of 10–11 years.

Respiratory Syncytial Virus (RSV) Infections

RSV contains a membrane fusion protein that induces fusion of infected cells with adjoining cells resulting in the formation of large multinucleated syncytia, from which the virus gets its name. It causes severe bronchiolitis in children (Table 4.21).

Morphology and replication is almost similar to PIV, and are transmitted by direct contact with secretions or fomites or by large droplet spread.

Clinical aspects   The illness starts like a common cold but within 24 hours the baby may be acutely ill with cyanosis and respiratory distress. Typically, there is bronchiolitis, with or without involvement of the lung parenchyma causing pneumonitis. Hypoxia and Apnea are common bindings of RSV infected children. The incubation period is about 5 days. There is a necrotizing broncholitis in which partial blocking of the bronchioles leads to the collapse of areas of lung.

Laboratory diagnosis   Detection of viral antigens by IF, ELISA of nasopharyngeal aspirates, and isolation of virus in cell cultures of LLC-MK2 (monkey kidney cell line) and HEp-2 cells. Characteristic syncytia are produced in 2–10 days. Detection of viral RNA using RT-PCR has also been used to determine RSV in clinical specimens.

Mumps

The name probably originates from an old word meaning 'tomope', an apt description for the miserable child afflicted by this common illness. Mumps was one of the first infections to be recognized and was described by Hippocrates as early as the 5th century BC.

Classified as member of the order *Mononegavirales*, family *Paramyxoviridae*, subfamily *Paramyxovirinae* and genus *Rubulavirus*. Virions are pleomorphic, spherical particles that range from 50 to 300 nm in diameter. An outer envelope encloses a helically coiled nucleocapsid. The envelope is about 10 nm thick and is composed of three layers. Glycoprotein spikes that project 10–15 nm, a lipid bilayer acquired from the host cell as the virus buds from the cytoplasmic membrane, and an inner structural matrix protein. The genome is a linear (–)-sense ssRNA of 15.3 kb that encodes six major proteins. Three proteins associated with RNP complex are the NP (nucleocapsid protein), P (phosphoprotein) and the L (large protein) protein (figure 4.31). The envelope contains the M protein and two surface glycoproteins that mediate HN and fusion (F) activities. NP is the major viral structural protein tightly encloses the genomic RNA.

Pathogenesis   Mumps is highly contagious, in humans particularly, near the time of onset of Parotitis. Mumps virus can be isolated from saliva from 5 days before to 5 days after the onset of clinical symptoms. Mean incubation period is 18days. After a primary replication at local site it spreads to regional lymph nodes, followed by transient plasma viraemia, resulting in dissemination of the virus to glandular and neural tissues. Virus can also be recovered from up to 72% of the urines collected during first 5 days of illness.

Table 4.21   Estimated frequency with which individual viral respiratory syndrome are caused by specific viral pathogens.

Virus	Frequency of syndrome					Pneumonia		
	Colds	Pharyngitis	Tracheobronchitis	Croup	Bronchiolitis	Children	Adult	Immuno compromised
RNA viruses								
Influenza A virus	+	++	+++	++	+	++	++++	+
Influenza B virus	+	++	++	+	+	+	++	+
PIV-1	+	++	+	++++	+	−	−	−
PIV-2	+	++	+	++	+	−	−	−
PIV-3	+	++	+	+++	++	+++	+	+
Respiratory Syncytial virus	++	+	−	++	++++	++++	+	++

Measles virus	-	-	+	+	-	++	+	+
Rhinovirus	++++	++	+	+	+	+	-	-
Enterovirus	++	++	-	-	+	+	-	-
Coronavirus	++	+	-	-	+	+	-	-
HIV	-	+	-	-	+	+	-	-
DNA viruses								
Adenovirus	-	++	+	++	++	++	++	++
Herpes simplex virus	-	+	-	-	+	+	-	+
Varicella virus	-	-	-	-	-	+	+	+
Epstein–Barr virus	-	++	-	-	-	-	-	-
Cytomegalovirus	-	+	-	-	-	++	-	++++

(-)-Occasional case report or rare; (+)—Cause some cases 1-5%; (++)—Fairly common cause 5-15%; (+++)—Common cause 15-25%; (++++)—Major cause >25% of cases.

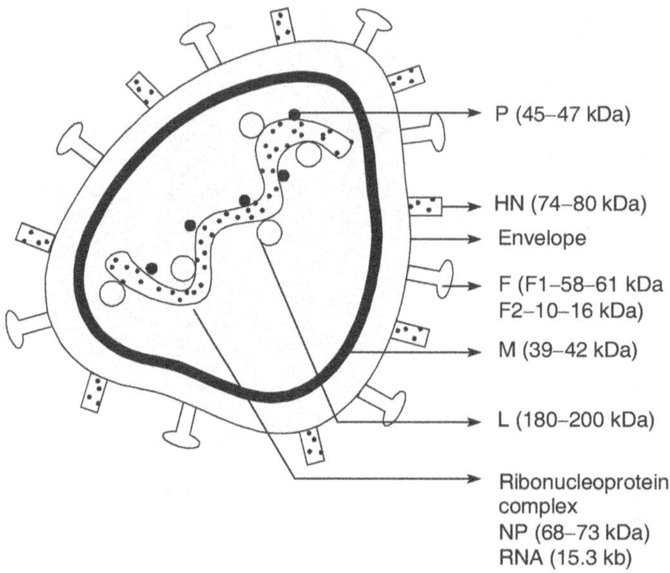

Figure 4.31   Structure of mumps virion molecular weights of viral protein in parentheses (as measured by gel electrophoresis) NP-Nucleocapsid protein; P-Phosphoprotein L - Large protein; HN - Hemagglutinin-neuraminidase; M - Matrix protein; F - Fusion protein.

Clinical aspects   The onset is marked by malaise and fever followed within 24 hours by a painful enlargement of one or both parotid glands; the other salivary glands are less often affected. In most cases, the swelling subsides within a few days and recovery is uneventful. Complications include Orchitis in about 20% of males who contract mumps after the age of puberty. Typically there is a swelling of one or both testicles 4–5 days after onset of parotitis. Incidence of aseptic meningitis is higher after mumps than after any other acute viral infection of childhood.

Laboratory diagnosis   Isolation of the virus from saliva, throat swab, CSF and urine on primary monkey kidney cells and HEp-2 cells, HAI and IF. Anti-mumps-IgM detection. ELISA is used for rapid diagnosis of mumps virus infection.

Prophylaxis A live attenuated vaccine given by subcutaneous injection in combination with attenuated measles and rubella virus strains—MMR vaccine to children of both sexes aged between 15–18 months.

Rubella

Rubella virus is the sole member of genus Rubivirus of the *Togaviridae* family. Rubella is derived from the Latin word *rubellus*, meaning 'reddish' and refers to the pink rash that is seen in most patients. Its popular name, German measles, probably derives from the fact that it was first described in Germany during the 18th Century. Rubella is predominantly an infection of children, in whom it causes a mild febrile illness.

The virus measure 50–70 nm in diameter. Virion is composed of 30 nm core surrounded by single layered envelope measuring 10 nm in thickness. Glycoprotein projections measure 5–6 nm in length on surface.

The genome contains ssRNA (40S) of positive sense of ~10,000 nucleotides, with a molecular weight of about 3.8×10^6. It has a capsid protein (C), two envelope glycoproteins E1 and E2 of 42 and 47 kDa respectively. E1 is viral haemagglutinin.

Rubella virus is heat labile. Rapid inactivation occurs at 56°C, infectivity is lost at temperatures between 10°C to –20°C. Chemicals such as ether, acetone, chloroform, formalin, beta-propiolactone, ethylene oxide 70% alcohol all inactivate virus. The virus is transmitted primarily by virus laden droplets from respiratory secretions of infected persons. It is observed from 7 days prior to 14 days following onset of the rash, with maximum shedding occurring from 5 days before to 6 days after appearance of exanthema. Incubation period is usually 16–18 days.

Foetal infection can occur throughout pregnancy. Rubella virus can cross placental barrier particularly during the first trimester of pregnancy and during term. Virus infects foetus, where it disseminates, grows and leads to congenital abnormalities or death of the foetus. The abnormalities include neurosensory deafness, cataracts, glaucoma, microphthalmia or retinopathy leading to blindness, congenital heart disease, microcephaly with mental retardation and hepatosplenomegaly. This is termed as congenital rubella syndrome (CRS).

Postnatal rubella   It is typically a mild disease which is acquired by inhalation. Virus multiplies in the upper respiratory tract and in cervical lymph nodes, then disseminates throughout the body by way of viraemia. Erythematous rash develops after an incubation period of 2–3 weeks with posterior cervical lymphadenopathy, fever, ophthalmalogia, conjunctivitis, malaise, anorexia, chills, cough, coryza and sore throat. Mild polyarthritis in hands is observed in females. The illness lasts of shorter duration and recovery is usually complete in about 3–4 days.

The main difference between rubella in children and in adults is that in the latter, polyarthritis is not an infrequent complication. It most often affects the hands and wrists, but may also involve the larger joints of the limbs.

Laboratory diagnosis   Virus isolation from throat swabs for 6 days before and after onset of rash by using AGMK cell lines. Serological diagnosis include 4-fold rise in rubella specific IgG between acute and convalescent serum. Demonstration of rubella specific IgM is indicative of recent infection. Rubella antigens can be detected by complement fixation test, haemagglutination inhibition (HAI), immunofluorescence (IF) and ELISAs.

Prevention   Active immunization includes RA27/3 rubella vaccine exclusively used in the US since 1979. Presently rubella vaccine is administered subcutaneously in combination with measles and mumps (MMR) at 15–18 months of age. A second MMR is then recommended at 11–12 years of age.

RHABDOVIRIDAE

Rhabdoviruses comprises of a family of (-) RNA virus and contains two genera which infect animals (Lyssavirus and Vesiculovirus). Type species of Lyssa virus is Rabies and that of Vesiculovirus is VSV (vesicular stomatitis virus).

Rabies

Properties   Rabies virus belongs to *Rhabdoviridae* (Greek, *rhabdos* = a rod), a family of characteristically bullet-shaped RNA viruses. Rabies and other viruses of the genus Lyssavirus (Greek, *lyssa* = madness) infect vertebrates. Rabies is enzootic and sometimes epizootic in a variety of mammal species includes wild and domestic canids (e.g. dogs, fox and coyotes), mustelids (skunks, badgers), viverids (mongooses, civets), procyonids (raccoons) and insectivorous and hematophagous bats.

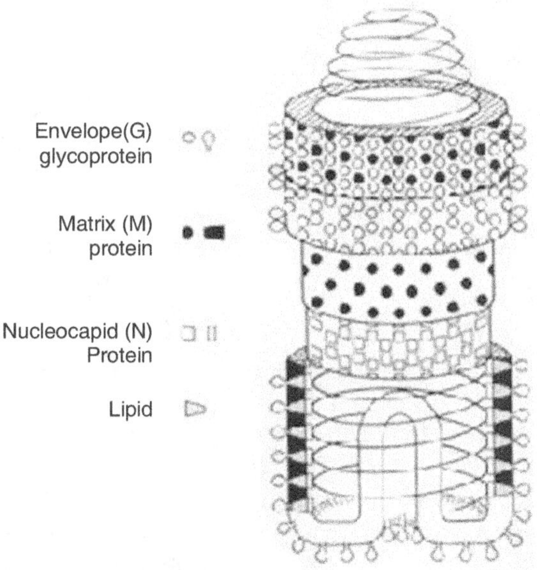

Figure 4.32   Rabies virus structure.

Capsid   Bullet shaped lipoprotein envelope with an average length of 180 nm and diameter at 75 nm, virion consist of a helical nucleocapsid with 30–35 coils. Protruding from the lipid envelope are ~200 glycoprotein (G) spikes of the virus constitute another 10 nm from envelope, covers entire coat except blunt end portion of envelope, responsible for viral attachment to cellular receptors. The M or Matrix protein is the major structural protein of the virus lying internally beneath the lipid membrane that encloses the genome (figure 4.32).

Genome   It consists of 11,932 nt length of non-segmented (-) RNA of 4.6 MDa. RNA has 5 genes, starting from 3' end they are named N, NS (or M1), M (or M2), G and L (Table 4.22).

Table 4.22   *Gene coding assignments of rabies virus.*

Gene	Synonym	Product size (kDa)	Function
N (nucleocapsid)		50	
NS (non-structural)	M1, P	40	Encode a structural protein which is phosphorylated (hence designated P) by host kinases and joins with L
M (matrix)	M2	26	
G (glycoprotein)		65	Attachment to host cell receptors
L (large)		160–190	RDRP, required for transcription of (-)RNA

Replication strategy   Nicotine acetylcholine receptor is one of the binding site for virus through which virus gets internalized by receptor mediated endocytosis. After fusion with lysosome, release of nucleocapsid in the cytosol occurs. Genome transcribe into 5' mRNA for production of viral protein. Then a full-length (+) strand is produced as template for the progeny viral genome. Viral envelope form from host cisternal membrane with which the G and M proteins are inserted (figure 4.33).

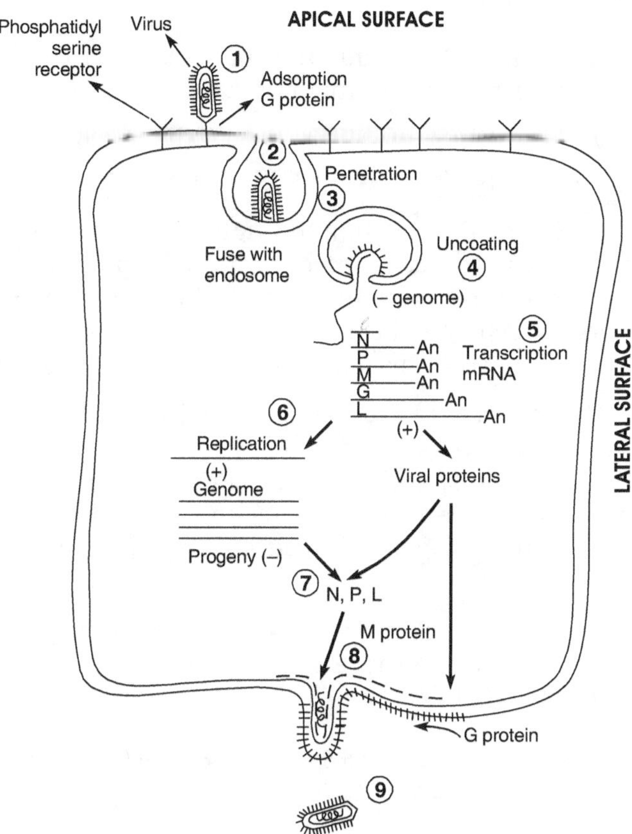

Figure 4.33   Steps in rhabdovirus infection: (1) virus attachment to phosphatidyl serine receptor (2) virus penetrates the cell in an endosome (3) virus fuses with endosomal membrane and core enters the cytoplasm (4) uncoating of nucleocapsid occurs (5) viral negative-sense RNA is transcribed into positive-sense RNA (6) positive-sense RNA serves as a template for synthesis of the viral genome as well as the mRNA that gives rise to viral proteins. This type of replicative process occurs with many viruses that form positive-sense RNA molecules (7) the negative-sense RNA becomes incorporated into nucleocapsids (N) (8) these NC subsequently join the matrix protein (M) at the basal surface of the cell (9) budding of virus from the cell surface takes place.

Pathogenesis

- Centripetal spread of virus via peripheral nerves (retrograde axoplasmic flow).
- Reaches the central nervous system.
- Centrifugal spread via peripheral nerves to many tissues.

Following introduction through a break in skin, replication occur in muscle cells infects muscle spindle infects nerve innervating the spindle, moves centrally through axons of these nerves and replicates in peripheral nerves. They appear in dorsal ganglion in 60–72 hr and are transported via sensory nerves.

Natural rabies infection may require a period of local viral replication, before CNS infection can occur. During this period administration of anti-rabies IgG and active immunization are able to prevent spread of virus to CNS. Adelchi Negri first described the cytoplasmic inclusions of rabies infection which bear his name 'Negri Bodies'.

Human rabies   The viral inoculum deposited during a bite and the location of the bite exerts an important influence on the likelihood of development of rabies (Table 4.23).

Table 4.23   Approximate mortality rates of patients after exposure to rabid dogs.

Location of exposure	Type	Extent	Mortality %
Face	Bite	Multiple, deep	60
Head (other than face)	Bite	Multiple, deep	50
Face	Bite	Single	30
Hand/finger	Bite	Severe	15
Face	Bite	Multiple, superficial	10
Hand	Bite	Multiple, superficial	5

Table 4.23 Contd.

Location of exposure	Type	Extent	Mortality %
Trunk/legs	Scratch	Superficial	3
Exposed skin	Kleeding	Superficial wound	2
Skin covered by clothing	Wound	Superficial	0.5
Recent wound	Contaminated by saliva		0.1

Duration of different stages of rabies

- Incubation period includes first 90 days (>75% cases).

- Prodrome and early symptoms: within 2–10 days duration followed by fever, headache, malaise, nausea, vomiting, paresthesia or pain at wound site.

- Acute neurologic disease: Human rabies infections are typically divided into two forms— furious and paralytic (or dumb). Furious rabies (80% cases) develop symptoms within 2–7 days present with hydrophobia, delirium and agitation, others include hallucination, bizarre behaviour, agitation, anxiety, biting, seizures, aerophobia, autonomic dysfunction and syndrome of antidiuretic hormone (SIADH).

- Paralytic "Dumb" rabies—patients lack hydrophobia, aerophobia and seizures. Their initial findings may suggest an ascending paralysis, resembling acute inflammatory polyneuropathy (Guillain–Barre syndrome), or a symmetric quadriparesis. Meningeal signs (headache, neck stiffness) may be prominent. The disease progresses to coma.

Growth characteristics

Street virus Freshly isolated virus from natural human or animal infection is termed street virus. It produces fatal encephalitis in laboratory animals, inoculated by any route, after long and variable incubation period of 1–12 weeks. Negri bodies can be demonstrated in the brains of animals dying of street virus infection.

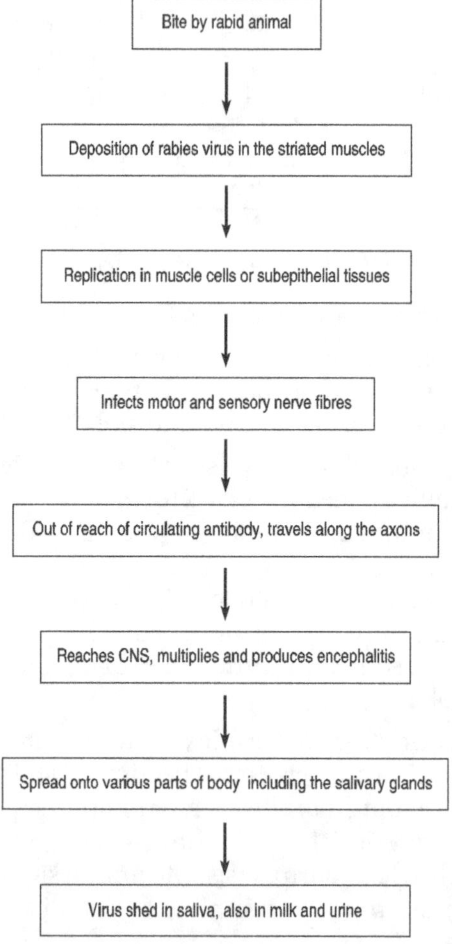

Fixed virus　Street virus after several serial intracerebral (i.c.) passages in rabbits, undergoes certain changes and becomes what is called fixed virus. It is more neurotropic and much less infective by other routes. After i.c. inoculation, it produces fatal encephalitis after a short and fixed incubation period of

5–8 days and Negri bodies are usually not demonstrable in the brain of animals dying of fixed virus infection. It produces paralytic rather than furious symptoms. It is used for vaccine production.

Culture   Rabies virus can be grown in several cell cultures including Baby hamster kidney (BHK), mouse neuroblastoma, human diploid lung fibroblasts, chick embryo fibroblasts and Vero monkey kidney cells with minimal cytopatheic effects. The last three are used in vaccine production.

Pathogenesis   Rabies virus is excreted in the saliva of infected animals of natural host (dogs, foxes, wolves, skunks, cats and bats). Humans acquire rabies virus by bite of rabid animals or following licks on abraded skin and mucosa by infected pet animals.

Laboratory diagnosis

1. *Demonstration of Negri bodies*   Sections or impression smears of brain stained by Seller's technique may reveal intracytoplasmic inclusion bodies, known as Negri bodies. They are oval or round, eosinophilic with basophilic inner granules of size range 3–27 mm in diameter seen in Purkinje cells of hippocampus, brain and cerebellum.

2. *Direct immunofluorescence detection of rabies antigen* Antemortem—Salivary, corneal or conjunctival smears or skin biopsy from the nape of the neck. Postmortem—in impression smears of the cut surface of the salivary gland, hippocampus, brain stem or cerebellum.

3. Detection of viral RNA and mRNA by RT-PCR assay and dot-blot hybridization assay on skin biopsy, corneal impression or saliva.

4. Isolation of virus in cell culture such as BHK, human diploid lung fibroblasts, Vero monkey kidney cells and detected by fluorescent antibody staining 18–24 hours post inoculation.

5. *Serology of Rabies*   Rabies antibodies in patients serum can be detected by ELISA.

Post-exposure treatment and prophylaxis

- Cleansing of wound with plenty of soap and water to eliminate virus load from the bite site.

- Human rabies immunoglobulin 20 IU/kg body weight to be administered intramuscularly in the gluteal region and second part to be infiltrated around bite. This is required in case of bite near head region.

- Rabies vaccine (culture derived) to be given on days 0, 3, 7, 14, 30 and 90 days in the deltoid region. Adequate tetanus prophylaxis should be ensured.

Rabies vaccines

Neural vaccines   Introduced by Pasteur in 1885 for post exposure prophylaxis, this contained grounded spinal cord of rabbit infected with rabies. This was dried for various length of period to inactivate the virus. Later on this was modified with inactivation procedures like rabies infected rabbit, sheep or goat brain subjected to virus inactivation with phenol at 22°C (Fermi vaccine), and at 37°C (Semple vaccine) or beta-propiolactone (BPL vaccines).

Non-neural vaccines

- *Duck egg vaccine*   Prepared from virus growth in duck embryos and inactivated with BPL were used for many years. It does not cause neuroparalytic accidents, but, a poor immunogen.

- *Cell culture vaccine*   Human diploid cell vaccines (HDCV) prepared from growth of fixed virus on human diploid lung fibroblasts and inactivated with BPL is widely used as it is highly immunogenic and free from side effects. Five to six doses of this vaccine given i.m. on the deltoid region on days 0, 3, 7, 14, 30 and 90 after exposure are recommended.

  Chick embryo cell vaccine and vero cell rabies vaccine of purified form have been developed to reduce the cost.

FLAVIVIRUSES

Yellow fever was the first human illness shown to be caused by a filterable virus in 1901 and isolated the yellow fever virus (YFV) in 1927 (Latin: *flavus,* meaning yellow and is derived from the name *Flaviviridae*). The virus causing Japanese encephalitis (JE), described clinically in the 1870s, was isolated from human brain tissue in 1935. Australian X-disease, an epidemic encephalitis illness described in 1911, shown to be caused by Murray valley encephalitis virus (MVE) in 1951. In the US, St. Louis encephalitis (SLE) was reported in 1920.

Dengue fever (DF) had been described in 18th century, isolated during world war-II, belong to group B arbovirus (now designated Flavivirus) and distinguished from group A arboviruses (now designated Alphavirus).

Flaviviruses are the most medically important group of arboviruses, DF and Dengue haemorrhagic fever (DHF) are major cause of human morbidity worldwide. YF remain an epidemic threat in Africa. The mosquito-borne and tick-borne encephalitis are responsible for endemic and epidemic disease in Americas, Europe and Asia. JE virus is the second most prevalent cause of viral encephalitis worldwide next to HIV.

Classification   The flaviviruses comprise a family of positive (+)-sense ssRNA viruses, majority of them are transmitted by arthropods (mosquitoes and ticks). Particles are composed of 40–60 nm in diameter, and have a spherical nucleocapsid surrounded by a lipid bilayer envelope with small surface projections, representing E glycoprotein dimers anchored onto the membrane and precursor prM protein that act as stabilizer for E protein. A pre M glycoprotein, present in intracellular nascent virions, is cleaved to the M protein found in mature extracellular virions. The genome is a ssRNA containing ~11 kb, contain a short 5´ noncoding region, a single long ORF and a 3´ noncoding terminus usually devoid of poly(A) tract. The ORF encodes 3 structural proteins at the 5´ end viz. capsid (C), pre-membrane (prM), and envelope (E) proteins. These are followed by 7 non-structural (NS) proteins (Table 4.24). Important biological activities are associated with the 53 kd E protein including haemagglutination, viral neutralization, virion assembly, membrane fusion and viral binding to cellular receptors.

Table 4.24   *Gene coding assignments of flaviviruses.*

Component	Designation	Molecular weight (kD)	Function
Structural	C (capsid)	11	Nucleocapsid structure/assembly, act as T-cell epitopes
	prM/M*	26/8	Stabilizes polymerizatio of E protein, neutralization antigen
	E	53	Major envelope protein, neutralization antigen, receptor binding, fusion haemagglutination, CTI target
Non-structural	NS1	24	Unknown
	NS2a	18	Unknown
	NS2b	13	Unknown
	NS3	69	Protease, RNA helicase, CTL
	NS4a	31	Replicase functions?
	NS4b	12	Replicase functions?
	NS5	104	RNA polymerase, methyltransferase

*prM cleaved during release from cell membrane, leaving M in virion envelope

Dengue Virus

Dengue is an important mosquito-borne disease in the world in terms of morbidity, mortality and economic costs, especially in the tropics, with more than 2/5th of the world's population living in areas are at risk for dengue. From being a sporadic illness, epidemics of dengue have now become a regular occurrence worldwide. It affects humans of all age groups in some parts of the world, it is mainly a paediatric public health problem. Dengue in infancy has been reported from Thailand, India and Sri Lanka.

Antigenic variation Four serotypes of Dengue virus have been reported. First description of Dengue haemorrhagic fever (DHF) and Dengue shock syndrome (DSS) was observed in 1956, and all the four types are implicated in DHF/DSS. Infection with an individual dengue virus confers lifelong protection against that serotype, but lack heterologous protection against other serotypes. Serotype-1 has 3 genotypes, serotype-2 has 6 genotypes, serotype-3 has 4 genotypes and type-4 has 1 genotype.

Transmission Principal vector mosquito that transmits dengue to humans includes *Aedes aegypti* in urban cycle and humans are the intermediate host. Period of viraemia during which humans are infectious for feeding mosquito (adult female) is 3–5 days. After feeding, an extrinsic incubation period of 10–14 days must elapse before *A. aegypti* can transmit the virus upon re-feeding.

Pathogenesis

Clinical manifestation The clinical features of dengue virus infection range from nonapparent infection through dengue fever (DF) and the more severe dengue haemorrhagic fever (DHF) and the dengue shock syndrome (DSS). The patients were diagnosed as having DF based on standard criteria like presentation of febrile illness of 2–7 days duration with features like headache, myalgia, arthralgia, rash, lumbosacral aching pain, retrobulbar pain, conjunctival congestion, puffiness of eyelids, haemorrhagic manifestations, leucopenia and deep bone pain, " breakbone fever" are characteristic features of the disease.

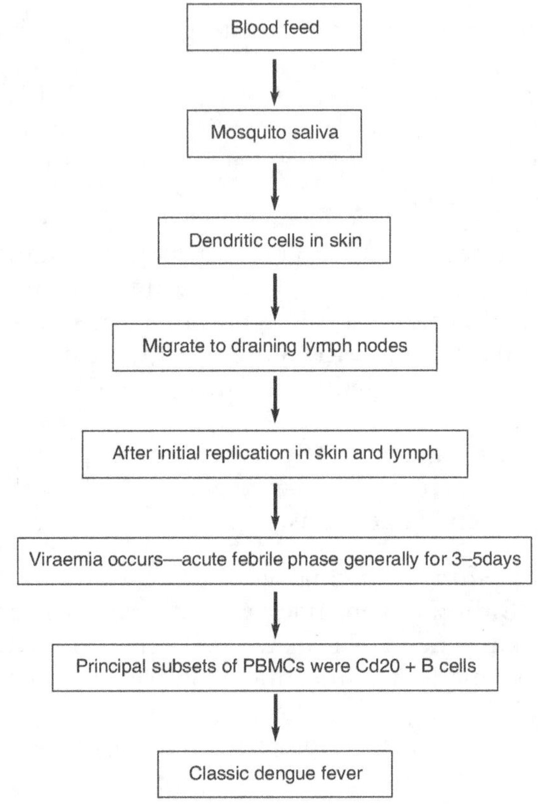

Laboratory diagnosis　Diagnosis may be made by isolation of the virus from blood during the first 3–5 days of illness. *Toxorhynchites* mosquitoes inoculated by intrathoracic injection are very sensitive hosts for isolation. Mosquito cell cultures *T. amboinensis* (TRA-284), *Aedes albolyticus* (C6/36), and *A. pseudoscutellaris* (AP-61) are highly susceptible to Dengue virus infection. Detection of the viral RNA by RT–PCR technique has been applied for rapid detection of Dengue virus. Dengue virus-specific IgM antibody tends to appear early during the course of disease and allows for a provisional diagnosis of dengue to be made from a single serum sample. Detection of dengue IgM antibodies is an easier method of diagnosing DF as compared to other classical serological methods like haemagglutination inhibition, neutralization and complement fixation tests. Serology includes 4-fold or higher rise in antibody titre. IgM capture ELISA is a useful diagnostic test.

Dengue virus infection rate in mosquitoes were determined by semi-nested RT-PCR. *A. aegypti* and *A. albopictus* were screened for Dengue virus. Two methods of RT-PCR, namely, 2-step nested RT-PCR and 1-tube multiplex RT-PCR, have been evaluated for rapid and cost-effective diagnostic test for detection as well as serotypic characterization of Dengue viruses in the acute phase of illness. One-tube multiplex RT-PCR was found superior; and its sensitivity and specificity were compared with gold standard "cell culture."

Prevention and treatment Prevention of Dengue transmission includes control or eradication of vector mosquitoes and elimination of breeding sites. Long term solution is the development of safe, effective, and inexpensive vaccines against all 4 serotypes. Studies have suggested that alpha-glucosidase inhibitors such as castanospermine and deoxynojirimycin inhibit dengue virus type 1 infection by disrupting the folding of the structural proteins prM and E, a step crucial to viral secretion. Recent reports documented that infections by all serotypes of Dengue virus were inhibited by castanospermine.

DHF Virus

DHF/DSS is an acute immunopathologic disease that occurs nearly in individuals sensitized by a prior infection and subsequently infected by a heterologous dengue virus serotype. Clinical features include hepatomegaly, leak of plasma from capillaries into extravascular space—hypotension. Case definition includes elevated haematocrit ($>20\%$ above normal), thrombocytopenia ($<100,000/cu.mm$) and positive tourniquet test. DSS diagnosis include criteria for DHF, along with reduction in pulse pressure (<20 mm Hg). All the four serotypes of Dengue virus cause DHF/DSS, but type-2 is most important followed by type-3.

Differential diagnosis   DHF is distinguished from DF by pressure of fever, thrombocytopenia (platelet count $<10^5$L) and haemoconcentration (haematocrit is more than 20%). DSS is a more severe form of the disease, characterized by hypotension (pulse pressure <20 mm Hg). Treatment includes O_2 administration, monitoring of blood pressure, pulse pressure, urinary output and vascular leak, measurement of the haematocrit and serum albumin level. Careful fluid management using Ringer's lactate or isotonic saline is required to treat cases of shock syndrome.

Dengue—Indian Scenario

The first isolation of dengue virus serotypes 1 and 4 was reported from India in 1964, and dengue virus serotype 3 in 1968. Ever since, intermittent reports of dengue and its sequelae have come from various parts of the country. These include reports from Ludhiana, Delhi, Lucknow, Calcutta, Chennai, Mangalore, Assam/Nagaland and Vellore. In India, dengue epidemics have been reported in many parts of the country. The incidence was reported to be high among children less than 8 years of age, and some infants presented with severe forms of dengue. Due to lack of awareness of dengue in general, the present surveillance system in India is unlikely to generate reliable dengue epidemiological database information, which is essential for better clinical assessment and management.

Between September 2001 and January 2002, an epidemic of dengue occurred in Chennai, Tamil Nadu, India. Nearly 800 cases were reported to the health system. According to World Health Organization clinical criteria, Dengue Fever (DF), Dengue Haemorrhagic Fever (DHF), and Dengue Hock Syndrome (DSS) were diagnosed in 82.8%, 3.4%, and 13.8% of serologically confirmed cases of Dengue in infants, respectively.

During the last few decades, dengue has reemerged in several parts of Southeast Asia, including India. A major outbreak of dengue infection occurred in northern India during October to December 2003. Virological and serological investigations confirmed that the outbreaks of fever were due to dengue virus infection. High breeding of *A. aegypti* in the study villages, detection of dengue-2 viral antigen and isolation of dengue-2 virus in *A. aegypti* mosquitoes confirmed the etiology reported in Kadumuchandiram and Mampatti villages of Dharmapuri district of Tamil Nadu, India.

From the high incidence of Dengue IgM seropositivity seen in 1999 and 2003, it appears that Dengue could probably be fast emerging as a major health concern in this part of India. Molecular studies on the circulating serotypes and their genotypes may be of help in addressing the probabilities of DSS/DHF incidence in future.

Japanese Encephalitis (JE) Virus

An outbreak of epidemic proportions occurred in the summer of 1924 in Japan and Korea. More than 6,000 cases, 60% fatal, were reported from Japan alone. Clinical and epidemiologic features of the illness suggest that this and subsequent outbreaks in 1927, 1934 and 1935 were epidemics of JE.

In 1924, a filterable agent from human brain tissue was isolated in rabbits and in 1934, Hayashi transmitted the disease experimentally to monkeys by intracerebral inoculation. JE viruses isolated from human cases in Japan in 1935 and in Beijing in 1949 provided the prototype Nakayama, Beijing and P3 strains respectively that are in principal use in vaccine production.

The virus initially was called *Japanese B encephalitis* (the modifying B has fallen into disuse) to distinguish the agent from the etiology of Von Economo's type A encephalitis, which had different epidemiologic characteristics. The mosquito-borne mode of JE transmission was elucidated with the isolation of JE virus from *Culex tritaeniorhynchus* mosquitoes in 1938 and in field studies that established the role of aquatic birds and pigs in the viral enzootic cycle.

Japanese encephalitis (JE) is a mosquito-borne arboviral disease of major public health importance in Asia. More than 35,000 cases and 10,000 deaths are reported annually from the region but official reports undoubtedly underestimate the true number of cases. First case of JE was reported in India during 1955. JE virus is the prototype of an antigenic complex that include viruses of SLE, MVE, West Nile virus, and other flaviviruses. Incidence rates are high in children less than 10 years and in the elderly. Transmission occurs by bites of infected mosquitoes, pigs and birds. Persistence of viraemia lasts several months after recovery from infection. Principal vectors are *Culicine* mosquitoes (*C. tritaeniorhyncus*) that use ground pools and rice paddies for their larval stages. Other important vectors include *C. fuscocephala*, *C. vishnui*, *C. bitaeniorhyncus* and *Anopheles hyrcanus* in India and Nepal.

After a mosquito bite, 4–14 days elapses before onset of symptoms. Vast majority of infections are cleared before neuroinvasion occur and are subclinical. Circulatory antibody plays a critical role in heterologous immunity.

Clinical symptoms   A prodromal illness of fever, headache, lethargy, nausea and vomiting lasting for several days precedes onset of CNS signs. Some develop seizures. Illness is fatal in 10–35% patients. It leads to foetal death and abortion if acquired during 1st and 2nd trimester of pregnancy. Generalized weakness and changes in tone, especially hypertonia and hyperreflexia, are the most common motor abnormalities, but focal motor deficits, including paresis and hemi- and tetraplegia, cranial nerve deficits (especially central facial palsy), and abnormal reflexes also may be present. Complication includes bacterial pneumonia.

Laboratory diagnosis   This includes screening of blood and CSF samples for virus detection by RT-PCR, and IgM capture ELISA. Specific IgM can be detected in CSF and or serum in approximately 75% of patients within the first 4 days after onset of illness and nearly all patients are positive 7 days after onset. In the majority of cases, IgM is detectable in CSF earlier than in serum; however, in some cases, the reverse holds and both fluids should be tested to maximize sensitivity.

Neuroimaging   MRI is superior to CT scan of Brain Cranial MRI reveals either mixed intensity or hypo intense lesion on T1 and hyperintense or mixed intensity lesion on T2 in thalami.

Thalamic changes may be helpful in the diagnosis of JE especially in endemic area.

Prevention

Vector Control   *C. tritaeniorhynchus* is the main JE vector in most areas of Asia. Although risk for acquiring JE is greatest in rural areas, enzootic viral transmission does exist in periurban areas allowing occurrence of sporadic JE cases. JE mosquito vectors are zoophilic; consequently, cows and certain other animals may reduce human risk by diverting vector mosquitoes (zoo prophylaxis). In India, JEV (Japanese encephalitis virus) has been isolated from 16 species of mosquitoes; the majority of the isolations were from *C. vishnui* complex, which breed extensively in rice ecosystem. The estimated JE morbidity rate reported from a hyperendemic area of Cuddalore district, Tamil Nadu vary from 0.3 to 1.5 cases per one lakh population.

Control of immature mosquitoes   Water management through intermittent irrigation has been adopted to control larvae of disease vectors in rice ecosystem. Composite culture of edible fish in the rice fields in Cauvery delta, Tamil Nadu has yielded promising feed back on the reduction of immature mosquito population and gain in the rice production and thus claimed to be an effective strategy in the control of JE vectors. Growing of the water fern *Azolla microphylla* in rice field was evaluated as a biological agent against mosquitoes breeding in rice fields.

Reduction in man–vector contact   Commercial mosquito repellants in the form of coil, cream and mats are widely practiced to repel mosquito bites. Pyrethroid–impregnated bed nets and curtains have been shown to reduce man–mosquito contact. A low cost technology was implemented on a pilot scale in JE-endemic areas of Cuddalore district. Pyrethroid impregnated jute/polypropylene curtains were used for protection against endophagic mosquitoes. The curtains were found to be effective and have proved its efficiency in the control of JE vectors.

Immunization against JE The other alternate left with for JE control is protection of the individual against JE by immunization. The three JE vaccines in widespread production and in worldwide use for this purpose are

1. Inactivated mouse brain-derived JE vaccine
2. Inactivated primary hamster kidney cell-derived JE vaccine
3. Live attenuated JE vaccine

Inactivated mouse brain–derived JE vaccine   The inactivated JE vaccines available for immunization purposes against JE are derived from infected mouse brain, which was licensed in Japan in 1954. This vaccine is produced by the Research Institute of Osaka University (Biken). The vaccine is prepared by subjecting the mouse brain infected with Nakayama strain of JE virus (JEV) to a sequence of protamine sulphate treatment, formalin inactivation, and ultrafiltration and ammonium sulphate precipitation. The vaccine is purified further by ultra centrifugation through a sucrose cushion. The purified product is without myelin basic protein and is supplied in a freeze-dried form. The Nakayama strain was chosen because of good propagation characteristics and has provided cross protection in mice against other JE-viral strains. This vaccine is manufactured in India, Japan, Korea, Taiwan, Thailand and Vietnam and in USA. Another JEV strain (Beijing-1) was also used in vaccine produced for domestic consumption by Biken, the principal Japanese manufacturer. Both the vaccines were found to be immunogenic in humans. Inactivated JE vaccine produced in mouse brain is commercially available.

Inactivated primary hamster kidney (PHK) cell–derived vaccine   In order to avoid brain antigens and allergic reactions associated with crude antigens and to ease production, tissue culture-derived vaccines were attempted in China. The supernatant collected from the virus infected cell line was inactivated with 0.5% formalin, stabilized with 0.1% human albumin and was tested for residual infectivity and potency. Both Nakayama and P1 and P3 (Beijing-1) strains were used and found to be immunogenic. Concentration of PHK-derived vaccine through ultra filtration followed by an ultracentrifugation (purified virion vaccine) was with higher potencies. The recommended dose was 1 ml given subcutaneously. Although this enhanced inactivated vaccine was reported to be immunogenic, there were no reports on efficacy or persistence of immunity.

Live attenuated vaccine   Attenuated JE viral strains have been generated through a number of passages in various cell culture systems including PHK cells. After 11 serial passages in suckling mice and 100 passages in PHK cells of SA14 JE strain, the neurovirulence of this virus in monkeys was lost. The final product titled as SA14-5-3 vaccine was shown to be safe and immunogenic. SA14-5-3 after a serial passage by subcutaneous inoculation of suckling mice, SA14-14-2 strain was obtained which was found to be immunogenic with stable attenuation. As a live vaccine, there are chances of transmitting the virus through the bite of a mosquito, since SA14-14-2 virus was present in the blood of the vaccine. Experimental infection of this virus in mosquitoes failed to transmit the virus. A dose of 0.5 ml is given subcutaneous at the age of one and two years. Efficacy trials in children 1–10 years old have yielded high protective rates.

Recombinant vaccines   Generations of vaccines through animal sources are not encouraged any longer due to objections from animal protection organizations for using animals for experimental purposes. To immunize people living in JE-endemic area of India, massive numbers of animals have to be sacrificed. Recent advances in molecular biology have made it possible to explore novel approaches for developing recombinant vaccines based on proteins, viruses and DNA.

Protein–based vaccines   Bacteria/viruses were used as expression systems to express genes coding for immunogenic proteins of JEV. Mammalian cell lines infected with recombinant viruses encoding JEV proteins were used as source to collect sub viral particles to be used as JE vaccine. Some of these derivatives were shown to be immunogenic inducing neutralizing antibodies in the animal hosts. The immune responses in the humans against these products remain to be investigated.

Virus-based vaccines   There are number of virus-based vaccines being evaluated in primates for safety and immunogenicity and protective efficacy. Two pox virus-vectored vaccines for JE, NYVAC-JEV and ALVAC-JEV were shown to be safe and immunogenic. NYVAC-JEV was effective in protecting monkeys against challenge with JEV. Live attenuated flaviviruses were developed against other flaviviruses. Chimeric flaviviruses were constructed in which the structural protein genes for the target antigens of a flavivirus were replaced by the corresponding genes of another flavivirus. The viral envelope genes of yellow fever 17D virus was replaced with the corresponding genes of JE SA14-14-2. The RNA transcripts of this chimera (YF/JE) were used to transfect Vero cells. The progeny virus was tested for neurovirulence in rhesus monkeys. Preclinical and clinical studies have shown that these vaccines were safe, immunogenic with protective efficacy.

DNA-based vaccines   A number of DNA vaccines are being tested for their immunogenicities, protective efficacy and safety. A JE vaccine candidate (pNJEME) encoding JEV premembrane (preM) and envelope genes was constructed in a plasmid vector and evaluated for safety and immunogenicity in cynomolgue monkeys. The vaccine induced neutralizing antibodies in the monkeys. It was demonstrated that pNJEME could induce JEV specific memory B cells. There were no side reactions. It was also demonstrated that a plasmid containing premembrane and envelope genes (pcDNA3JEME) induced JEV-specific memory cytotoxic and B cells in mice. In recent years a number of efforts are in progress to develop a molecular based JE vaccine. The vaccines developed were tested for safety and efficacy in animal models and claimed to be immunogenic and safe. Though, the results from the animal experiments are encouraging, trails in humans will deliver the net outcome of these vaccines.

Treatment   Ribavirin is active in vitro and no specific antiviral therapy is presently available.

JE *Indian scenario*   Japanese Encephalitis, popularly called "Brain Fever", is caused by a virus. Since it is carried by mosquito, an arthropod, it is classified under Arbovirus. Indian strain of Japanese encephalitis virus (JEV) is GP78, which is phylogenetically closer to the Chinese SA14 isolate.

In India, JE was first recorded in Vellore in Tamil Nadu and Pondicherry in mid 1950s. JE has been reported from 24 states/ Union territories so far. Frequently affected states include Andhra Pradesh, Assam, Bihar, Goa, Haryana, Karnataka, Manipur, Tamil Nadu, Uttar Pradesh and West Bengal. The Directorate of National Anti Malaria Program (NAMP) is monitoring JE in India since 1978. An estimated 378 million population is living at the risk of JE in 12 states/Union territories that are frequently affected. The spread of JE to new areas is probably due to agricultural development and intensive rice cultivation supported by irrigation schemes.

It has high mortality and morbidity rates. Usually affected age group is 5–10 years though children from 3–14 years can be affected. In West Bengal, only 36% were children and 64% were adults probably due to fresh introduction of virus. This trend was seen in Korea also, when paediatric population received JE vaccine.

West Nile virus—the Indian scenario   West Nile virus (WNV) is an important arthropod borne flavivirus; usually causes a mild infection called West Nile fever (WNF) in humans and horses. Mosquitoes are the principal vectors of WNV. Various Culex species are found to act as vectors in different geographical regions. The virus is maintained in a bird-mosquito cycle in nature. In India, Culex mosquitoes are tentatively incriminated as vectors of WNV. Experimental studies have shown that *Culex tritaeniorhynchus, C. vishnui, C. bitaeniorhynchus* and *C. univittatus, Cx. pipiens fatigans* and *A. albopictus* could act as potential vectors of WNV. Transovarial transmission of WNV has been experimentally demonstrated in Culex mosquitoes. Apart from mosquitoes, the role of other arthropods is also considered in the maintenance of WNV during inter-enzootic periods. The possible role of ardeid birds in the maintenance of WNV has been described in India. Though very few clinically overt cases of human encephalitis due to WNV are observed, JEV is found to dominate in southern India. WNF in horses has not been documented in India. JEV immunized monkeys were protected from WNV challenge and the WNV immunization was found to reduce the disease severity due to JEV. Based on the limited genome sequence analysis, the Indian isolates are grouped together under the genetic lineage-I. WNV infection is diagnosed by IgM antibody capture enzyme linked immunosorbent assay, haemagglutination inhibition test, neutralization test and reverse transcriptase–polymerase chain reaction (RT–PCR). For the effective control of Culex mosquitoes, integrated vector control strategies are recommended. Specific methods are not available for the treatment of WNV infection. However, in patients with encephalitis supportive therapy is recommended. Though a few candidate vaccines are under laboratory trial, no vaccine has been available commercially for the control of WNV infection in human and animals.

Serosurveillance   The following institutions are identified for Serosurveillance.

1. National Institute of Virology, Pune

2. NICD (National Institute of Infectious Diseases), Delhi

3. School of Tropical Medicine, Calcutta

4. Centre for Research in Medical Entomology, Madurai

5. KG Medical College, Lucknow

6. Gorakhpur Medical College, Gorakhpur

7. Kings Institute of Preventive Medicine, Chennai

8. Burdwan Medical College, Burdwan

9. Assam Medical College, Dibrugarh

10. VBRI (Veterinary Biological Research Institute), Shanthinagar, Hyderabad, Andhra Pradesh

11. Kyasanoor Forest Disease Laboratory, Shimoga, Karnataka

12. Institute of Vector control and Zoonosis, Hosur, Tamil Nadu

13. Central Research Institute, Kasauli

14. Goa Medical College, Panaji.

Hepatitis C Virus (HCV)

HCV is an enveloped virus measuring 55 to 65 nm in diameter with spike like projections. The genome contains ssRNA of (+)-sense with 9 dkb. All the viral proteins are thought to be produced from the largest ORF in the genome. One of the factors that hampered the isolation of HCV was thought to be the low level of HCV replication in infected human hosts. However, isolation of HCV cDNAs from patients became routine using PCR.

Based on the nucleotide sequences of HCV nine major genotypes have been suggested and designated as HCV-1 to HCV-9. Some of these types contain several closely related subtypes designated with a lower case letter following the type number (HCV-1a, -1b, HCV-2a, -2b, and so forth).

Viral proteins   HCV is comprised of the following proteins

- Nucleocapsid protein that bind RNA (core or C; 17–22 kDa)
- 2 putative virus envelope glycoproteins (E1; 33–35 kDa and E2; 68–72 kDa).
- E2 was once called non-structural protein 1 (NS1)
- HCV non-structural polyprotein gives rise to 6 NSPs, NS2, NS3, NS4A, NS4B, NS5A and NS5B.

Replication   The HCV viral genome comprising of (+)-sense RNA function as an mRNA in infected cells. Translation of an HCV polyprotein proceeds in an internal ribosome entry site (IRES) dependent manner that is cap independent. During replication HCV synthesizes (–)-sense RNA to use as a template for generating progeny virion RNAs.

Transmission   Transmission through blood and blood products seems to be the major route of HCV infection. Several percutaneous, possibly natural routes of HCV transmission include sexual, perinatal, and intrafamilial transmission (Table 4.8a).

Pathogenesis   HCV infection causes acute and chronic hepatitis. Acute hepatitis is followed by chronic state in ~50% of cases and 20% of chronic hepatitis cases will develop cirrhosis and hepatocellular carcinoma (HCC). Incubation period is about 5 to 12 weeks. Presence of HCV antibody and HCC are strongly associated. HCV infection is associated with a high proportion of patients who have fulminant hepatitis.

HCV infection does not elicit protective immunity against reinfection with homologous or heterologous strains. HCV RNA can be detected in serum, liver biopsy, circulating monocytes, and body fluids. Negative (−) sense RNA intermediate of HCV is an indicator of virus replication in hepatocytes. All chronic active hepatitis cases are antibody positive for HCV anticore or anti-NS4A protein, and HCV RNA.

Clinical manifestation

Acute hepatitis   The acute disease often goes unrecognized at least 2/3rd of the cases are anicteric or completely asymptomatic. Incubation period varies from about 2 weeks to 6 months with an average of 7–8 weeks.

Chronic hepatitis   Chronic hepatitis is the leading cause of liver transplantation due to chronic liver failure. Chronic HCV infection leads to cirrhosis, chronic liver failure and HCC. It leads to slower progression of the disease to take about 10 to 21 years after transfusion, cirrhosis appears 21 to 30 years after transfusion, and HCC 29 to 42 years after transfusion. Detection of HCV RNA by PCR remain authentic.

Laboratory diagnosis

1.   HCV RNA detection by RT-PCR

2.   HCV core antigen detection by enzyme immuno assay

3.   Anti-core protein antibody to HCV

4.   Anti-HCV-IgM to core protein

5.   Detection and quantitation of HCV RNA by branched chain DNA (bDNA) technology, which is based on hybridization with branched oligonucleotide that contains a unique primary segment with complementarity to an HCV sequence; a set of secondary fragments with multiple copies

of a small alkaline phosphatase labeled oligonucleotide is covalently linked to the primary sequence through the branched points.

FILOVIRIDAE

Ebola Virus

The emerging pathogens Marburg and Ebola cause severe haemorrhagic fevers with mortalities as high as 90%, Ebola outbreak first recognized in 1976 in Africa, and Marburg virus first isolated in 1967 remains elusive of their sources. These viruses are placed under the genus *Filovirus* (Latin *filum* = thread), and have genomes of (–)-sense RNA and a distinctive filamentous morphology. Marburg virus has no known subtypes, whereas Ebola virus has at least four subtypes (Zaire, Sudan, Ivory Coast, and Reston). There is no antigenic cross reactivity between Marburg and Ebola viruses. The particles are of uniform 80 nm diameter with a mean unit length of 1250 nm in case of Ebola virus (figure 4.34) and 865 nm in case of Marburg virus.

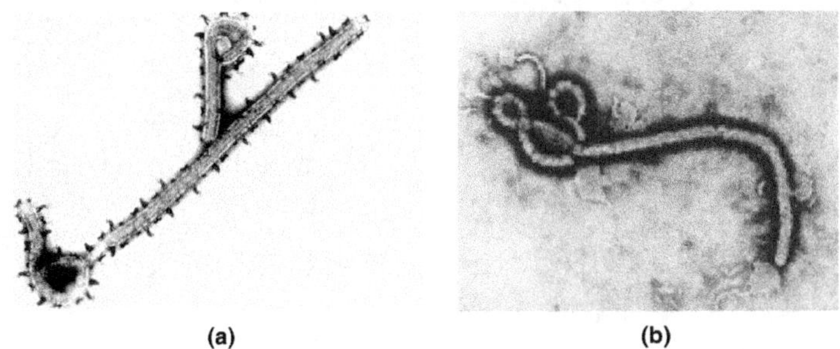

Figure 4.34   Ebola virus (a) Reston type (b) Zaire type.

The genome of (–) RNA is of ~19 kb in length consisting of seven sequentially arranged genes. The genome has the following characteristics:

- long non-coding sequence at 3´ and 5´ end.

- conserved 'UAAUU' transcriptional signal at 5´ end of start site and 3' end of stop site.

- There are three overlaps in the Zaire subtype of Ebola virus that alternate along the genome with intergenic regions (VP30 – VP24) as shown in figure 4.35.

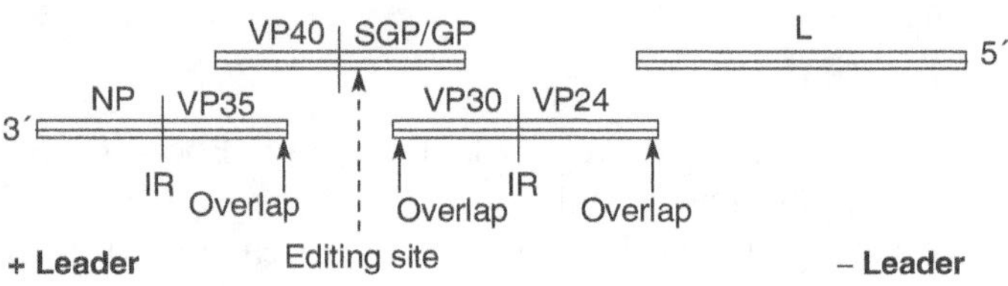

Figure 4.35 Schematic representation of the genome of Ebola (19.0 kb) IR - Intergenic region; L - Polymerase protein (large) sGP - Secreted glycoprotein; NP - nucleoprotein.

Viral proteins The types and functions of different virus proteins are stated in Table 4.25. and the salient structural composition is as follows:

- The virion surface spikes of 8–10 nm are composed of a single structural protein.

- Central ribonucleoprotein core is covered by lipid envelope.

- Glycoprotein (GP) on the surface and two matrix proteins (VP40 and VP24) that will connect lipid bilayer and GP to nucleocapsid protein.

- Two nucleoproteins (NP and VP30 encase the genomic RNA).

Clinical Manifestation

Following an incubation period of 4 to 10 days, abrupt onset of illness occurs, with nonspecific symptoms including fever, severe frontal headache, malaise and myalgia. Early signs include bradycardia and conjunctivitis. After 2–3 days of onset pharyngitis, severe nausea, and vomiting, progressing to hematemeses and melena occurs. Bleeding is the significant factor and manifests as petechiae, echymoses, uncontrolled bleeding from venipuncture sites, and postmortem evidence of visceral haemorrhagic effusions. Macropapular rash appear on day 5 usually followed by desquamation in survivors. Convalescence is slow, often takes 5 weeks or more, marked by prostration, weight loss, and amnesia for the period of acute illness.

Table 4.25   Filovirus proteins.

Designation	Type	Function	Size (in kDa)
Glycoprotein	Glycoprotein	Virion surface spike	120–170
VP40	Protein	Analogy to matrix protein, facilitate budding of virion at plasma membrane surface	40
VP24	Protein	Membrane protein	24
NP (nucleoprotein)	Protein	Major nucleocapsid	96–104
VP30	Protein	Minor nucleocapsid	30
L protein	Large Protein	RNA-dependent RNA polymerase	
VP35	Glycoprotein	Possible role in replication	35
Secreted glycoprotein (SGP)	Non-structural	Secreted form of glycoprotein	

Laboratory Diagnosis

Because of their aerosol infectivity, high mortality rate, and potential for person-to-person transmission, filoviruses are classified as Biosafety level 4 (BSL-4) pathogens by the World Health Organization Risk Group 4, for which maximum containment facilities are required for all laboratory work. The diagnosis include the following:

- Isolation of the virus in Veroclone E-6 cells
- Electron Microscopic detection of virus particles by negative staining
- Antigen detection by ELISA
- Serology include direct and indirect IgG and IgM ELISA tests.

BUNYAVIRIDAE

The family *Bunyaviridae* is the largest family of animal viruses, including many that are known human pathogens. The clinical diseases produced in humans range from acute febrile illnesses, such as sandfly fever, to more distinct clinical syndromes such as California encephalitis, Rift Valley fever, Crimean Congo haemorrhagic fever (CCHF), haemorrhagic fever with renal syndrome (HFRS), and the recently recognized Hantaviruses pulmonary syndrome (HPS). The family *Bunyaviridae* is divided into five genera: *Bunyavirus, Phlebovirus, Nairovirus, Hantavirus,* and *Tospovirus.* All the genera except the tospoviruses, which are plant viruses, infect vertebrate hosts and include human pathogens. There are more than 300 bunyaviruses in the family, and most but not all are assigned to one of the five genera. Representative groups and complexes for the *Bunyaviruses, Phleboviruses, Nairoviruses,* and *Hantaviruses,* including major human pathogens, are shown in Table 4.26.

HANTAAN VIRUS

Virion Structure

Bunyaviruses are enveloped viruses about 100 nm in diameter with 5 to 10 nm glycoprotein spikes projecting from the envelope (figure 4.36). The viruses are generally spherical and the envelope surrounds a core that consists of three nucleocapsids comprising the short (S), medium (M), and long (L) genome segments respectively. The genome is (–)-sense ssRNA. The length of the genomic segments vary among the genera, ranging from 6.3 to 12 kb for the L segment, 3.5 to 6 kb for the M segment, and 1 to 2.2 kb for the S segment. For Hantaviruses, segment lengths range from 6.5 to 7 kb for the L segment, 3.6 to 3.7 kb for the M segment, and 1.6 to 2.1 kb for the S segment.

Table 4.26   *Bunyaviridae* genera, subgroups, and selected viruses that are human pathogens.

Genus	Serogroup of Complex	Geographic Location of Human Pathogens	Selected Human Pathogens
Bunyavirus	California Simbu Anopheles A Bunyamwera Group C	North America, Europe South America South America Africa, South America	California encephalitis, LaCrosse, Oropouche virus Tacaiuma virus Bunyamwera, Fort Sherman
		North and South America	Apue, Itaqui, Madrid, Oriboca viruses
Nairovirus	Crimean-Congo hemorrhagic fever (CCHF)	Africa, Asia, Europe	CCHF virus
	Nairobi sheep disease	Africa, Asia	Dugbe, Nairobi sheep disease viruses
Phlebovirus	Sandfly fever group	North and South America, Africa, Asia, Europe	Alenquer, Punta Toro, Rift Valley fever, sandfly fever, Naples
Hantavirus	Haemorrhagic fever group	Japan, Korea, Russia, USA, China, Scandinavia	Hantaan, Puumala, Muerto Canyon

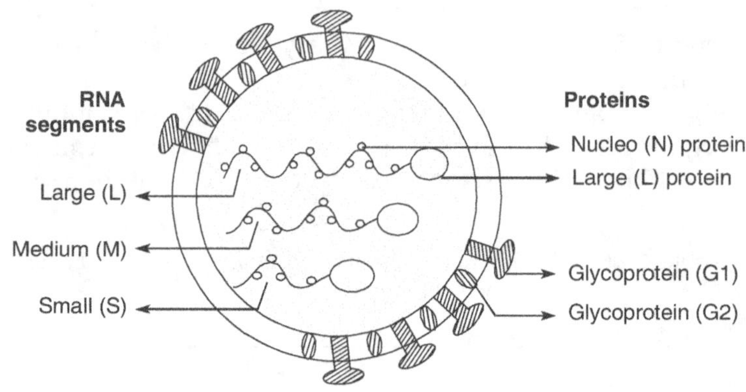

Figure 4.36   Structure of a bunyavirion.

The L Protein plays a role in both transcription and replication. The 6,530–nucleotide sequence of the Hantaan L segment codes for a polypeptide of 246.5 kDa, and the M segment codes for the glycoproteins, G1 and G2, and for a nonstructural protein NS_m), although the latter has not been detected in the hantaviruses. The S segment codes for the nucleocapsid protein (N) and for a nonstructural protein (NS_s), although the nonstructural protein also has not been detected in the hantaviruses.

Hantaviruses infect only vertebrates and have no arthropod hosts. Although "spillover" may occur between small rodent species in nature and although rats, mice , and rabbits can be infected in the laboratory , each hantaviruses tends to have a highly restricted host range in nature (rodent - *Apodesmus agrarius*)

Hantaviruses Pulmonary Syndrome (HPS)

Hantavirus pulmonary syndrome (HPS) is a deadly disease transmitted by infected rodents through urine, droppings, or saliva. Humans can contract the disease when they breathe in aerosolized virus. Although rare, HPS is potentially deadly. Rodent control in and around the home remains the primary strategy for preventing hantavirus infection.

HPS was first recognized in May 1993 following a cluster of unexplained deaths in young adults in the rural, Four Corners area of northwestern New Mexico and north – eastern Arizona. A previously unrecognized hantavirus associated with the deer mouse, *Peromyscus maniculatus*, was found to be the virus responsible for most cases of HPS, and specific serologic tests and tests for nucleic acid detection (reverse transcriptase-polymerase chain reaction; RT–PCR)

Early Symptoms

Early symptoms include fatigue, fever and muscle aches, especially in the large muscle groups-thighs, hips, back, and sometimes shoulders. These symptoms are universal. There may also be headaches, dizziness, chills, and abdominal problems, such as nausea, vomiting, diarrhea, and abdominal pain. About half of all HPS patients experience these symptoms.

Late Symptoms

Four to 10 days after the initial phase of illness, the late symptoms of HPS appear. These include coughing and shortness of breath, with the sensation of, as one survivor put it, a "...tight band around my chest and a pillow over my face" as the lungs fill with fluid.

Uncommon Symptoms

Ear ache, sore throat, runny nose, and rash are very uncommon symptoms of HPS. On the basis of limited information, it appears that symptoms may develop between 1 and 5 weeks after exposure to urine, droppings, or saliva of infected rodents.

There is no specific treatment, cure, or vaccine for hantavirus infection. However, we do know that if infected individuals are recognized early and receive medical care in an intensive care unit, they may do better. In intensive care, patients are intubated and given oxygen therapy to help them through the period of severe respiratory distress.

Hemorrhagic Fever with Renal Syndrome (HFRS)

HFRS associated with Hantaan virus (HTN) and *Apodesmus agrarius* (striped field mouse) is distributed throughout the Far East , particularly in China and Korea, eastern Russia, and the Balkans. Although infection with the hantaviruses causing HFRS occurs in all age groups, infection and disease peak in adults 15 to 40 years of age; like HPS, HFRS is uncommon in children.

Pathogenesis in Humans

The incubation period between rodent contact and onset of symptoms has often been difficult to determine, since the period of potential exposure may be prolonged. A small study of persons with nephropathia epidemica suggested that the median was about 4 weeks, with a range of 1 to 8 weeks. For the more severe forms of HFRS, the average incubation period has been estimated to be 12 to 21 days , with a range of 4 days to 6 weeks. Information on the incubation period for HPS is even more limited but has been estimated to be 8 to 21 days.

Replication

Hantaviruses replicate exclusively in the host cell cytoplasm. Entry into host cells is thought to occur by attachment of virions to cellular receptors and subsequent endocytosis. Nucleocapsids are introduced into the cytoplasm by pH–dependent fusion of the virion with the endosomal membrane. Transcription of viral genes is initiated by association of the L protein with the three nucleocapsid species. In addition to transcriptase and replicase functions, the viral L protein is also thought to have an endonuclease activity that cleaves cellular messenger RNAs (mRNAs) for the production of capped primers used to initiate transcription of viral mRNAs. As a result of this "cap snatching," the mRNAs of hantaviruses are believed to be capped and contain non–templated 5´ terminal extensions. The viral N and L mRNAs are thought to undergo translation at free ribosomes, whereas the M mRNA is translated in the endoplasmic reticulum. G1 and G2 glycoproteins form heterodimers and are then transported from the endoplasmic reticulum to the Golgi complex, where glycosylation is completed. The L protein produces nascent genomes by replication via a positive-sense RNA intermediate. Hantavirions are believed to form by association of nucleocapsids with glycoproteins embedded in the membranes of the Golgi, followed by budding into the Golgi cisternae. Nascent virions are then transported in secretory vesicles to the plasma membrane and released by exocytosis.

Clinical Manifestation

Hantavirus pulmonary syndrome   HPS typically begins with a febrile phase characterized by fever and myalgia ; the latter is often prominent. Headache, backache, abdominal pain, diarrhoea, nausea, and vomiting may also be present, particularly after the first 24 to 48 hours. The febrile phase which usually lasts for 3 to 4 days is followed by the sudden onset of noncardiogenic pulmonary edema and, in some cases, by shock. The shock phase is heralded by abrupt onset of cough and shortness of breath, which may be accompanied by dizziness. Pulmonary edema develops rapidly, and patients become hypoxic at this stage, and nearly all have required supplemental oxygen.

Laboratory diagnosis   Growth in cell culture using Vero E6 cells, Virus specific IgM and IgG in cases of HFRS. Available serologic tests include ELISA, neutralization assay, HI, western immunoblot, recombinant immunoblot assay (RIBA). Reverse transcriptase–PCR (RT–PCR) can be used to detect hantaviral RNA in fresh frozen lung tissue, blood clots, or nucleated blood cells.

REVIEW QUESTIONS

1.   Write short notes on:
     a.   Antigenic shift and antigenic drift
     b.   Fixed and street virus
     c.   The first human virus identified

    d. Negri bodies

    e. Shingles

    f. Episomes

    g. Burkitt's Lymphoma

    h. HCC

    i. HBsAg

    j. Coryza

    k. German measles

    l. Erythema infectiosum

    m. Genital warts

    n. HBeAg

    o. Rotavirus

    p. NACO

    q. Hairy leukoplakia

    r. AIDS Dementia complex

    s. Hemagglutinin and Neuraminidase

    t. Koplik spots

    u. Syncytia

    v. Yellow fever virus

    w. DHF

    x. Cytopatheic effect

2. Draw the diagram of HIV virus.

3. Give an account of morphology, replication, clinical features and laboratory diagnosis of HCV.

4. What are the serologic markers of Hepatitis B virus infection?

5. Write short notes on inclusion bodies.

6. Give an account of the classification of animal viruses.

7. Write a note on the replication of dsDNA viruses.

8. What are the cultivation methods of viruses?

9. Write a short note on PCR.

10. What are the laboratory methods in diagnosis of virus infections?

11. Discuss the epidemiology, diagnosis, pathogenesis, prevention and treatment of poliomyelitis infection.

12. Give a note on viruses of CNS and add a note on arboviruses.

13. Discuss the epidemiology, characteristics, life cycle, pathogenicity and clinical manifestations of Epstein-Barr virus.

14. Characteristics, pathogenesis, clinical features and laboratory diagnosis of HSV-2 infection.

15. Describe the antigenic structure, replication, modes of transmission and prevention and treatment of HIV infection

16. Give a note on Infectious mononucleosis

17. Clinical syndromes and laboratory diagnosis of Adenovirus infections

18. Give an account of the pathogenesis, laboratory diagnosis, post-exposure treatment and prophylaxis of rabies virus.

19. Write a detailed note on the replication mechanism and laboratory diagnosis of retrovirus HTLV.

20. Give an account of the morphology and characteristics of Hantaan virus, and detail on Hantavirus pulmonary syndrome.

21. Give a detailed description of influenza A virus antigenic structure, replication, pathogenesis and lab diagnosis. Add a note on its treatment and prophylaxis.

22. Give a detailed note on the morphology, replication, clinical features and diagnosis of smallpox virus.

5 ANTIVIRAL CHEMOTHERAPY AND VIRAL VACCINES

Nowadays, vaccines have occupied the central position in attempts to control virus infections. They are relatively cheap and safe and confer long-term immunity. However, some viruses, for some reasons, are not fully amenable to this approach, such as influenza virus, retroviruses, herpesviruses, the slow viruses, rhinoviruses and arboviruses. Obstacles to the use of vaccines include:

a. multiplicity of serotypes, e.g. rhinoviruses, togaviruses

b. antigenic change, e.g. influenza virus, retroviruses

c. latent infections

No antiviral compound tested has been able to inhibit completely the replication of any virus and a proportion of viral particles always seem to be able to circumvent the drug-induced blockade.

THE CHEMISTRY OF ANTIVIRAL COMPOUNDS

There are few restrictions on the types of molecules that inhibit virus replication, at least in the laboratory. They vary greatly in complexity and include natural products of plants, synthetic oligonucleotides, oligosaccharides, simple inorganic and organic compounds and nucleoside analogues. Examples of antiviral compounds in current use include:

Nucleoside analogues   Thousands of analogues of naturally occurring nucleosides have now been synthesized and tested in the laboratory, initially as herpesvirus inhibitors and nowadays many are retested as anti-HIV agents. In addition to purine and pyrimidine nucleosides, ara-, amino-, aza-nucleosides or nucleotides have also been synthesized. Even single atomic substitution may change an active molecule to an inactive one.

Non-nucleoside analogues   These comprise a number of compound classes, including the dipyridodiazepinones, HEPT (1-[2-hydroxyethoxymethyl]-6-[phenylthio]thymines), bis (heteroaryl)piperazines (BHAP), phenylethylthioureathiazole (PETT) derivatives, benzoxazinone derivatives, and other novel agents.

Pyrophosphate analogues Foscarnet is an example of a pyrophosphate analogue. This specifically inhibits herpesvirus DNA polymerase at the pyrophosphate binding sites and it also has anti-HIV activity.

Amantidine molecules   Amantidine is licensed for the treatment of influenza A infection. Addition of a methyl grouping (rimantidine) alters the pharmacological distribution of the drug and prevents entry to the brain, thus reducing the side effect described as "jitteriness."

COMMONLY USED ANTIVIRAL AGENTS

Some commonly used antiviral agents shown in Table 5.1 are discussed below.

Acyclovir

Acyclovir is a synthetic guanine nucleoside analogue (figure 5.1, Table 5.1). The initial step of phosphorylation to ACV monophosphate is preferentially carried out by viral thymidine kinase rather than cellular kinases. The monophosphate cannot leave infected cells so that more non-phosphorylated compounds enter to make up for the depleted intracellular concentration, only to be converted to the monophosphate. In this manner, the drug accumulates in the herpes-infected cells rather than in the uninfected counterparts. The monophosphate is then

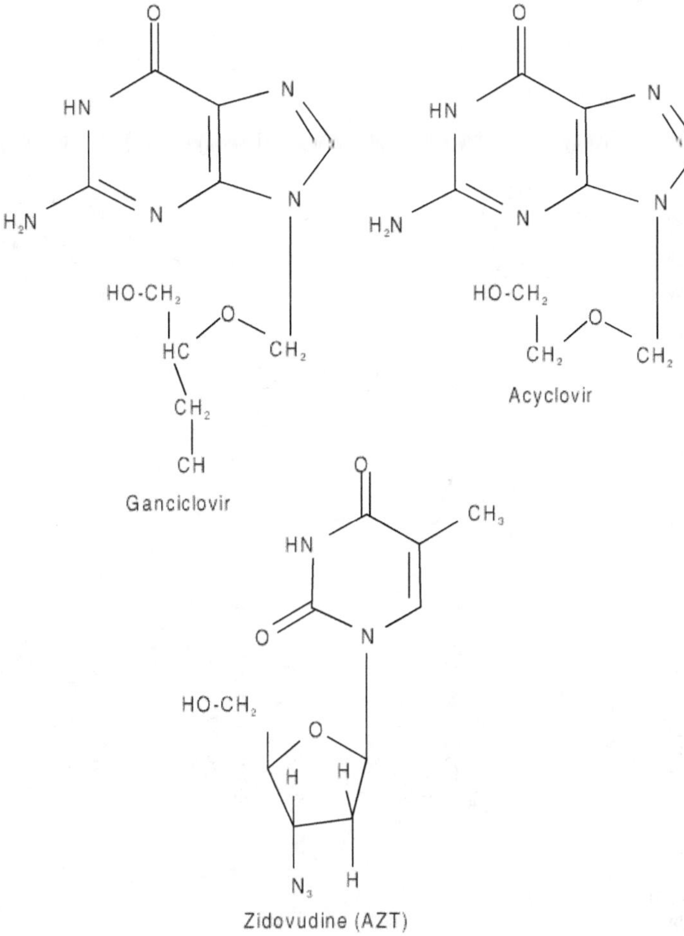

Ganciclovir

Acyclovir

Zidovudine (AZT)

Figure 5.1   Chemical structures of few anti-herpes and anti-retroviral agents.

phosphorylated to the di- and triphosphate forms by cellular enzymes. ACV triphosphate is the pharmacologically active form of the drug. It inhibits herpes DNA polymerase with little effect on the host cell DNA polymerase. It also has some chain termination activity and thereby it behaves as a "suicide inhibitor."

Acyclovir-resistant strains of HSV have mutations in either the viral thymidine kinase gene or the viral DNA polymerase.

Table 5.1   Antiviral agents—Indications and dosing regimens.

Antiviral agent	Property	Mechanism of action	Indication	Dosage
Anti–herpesviral agents				
Vidarabine	Nucleoside purine analogue	Undergoes phosphorylation to form triphosphate that acts as a competitive inhibitor of viral DNA polymerase	Herpes simplex keratitis systemic use	Apply to conjunctivae 5id 15mg/kg I.V. qd
Acyclovir	Guanosine analogue	Phosphorylated form of acyclovir (triphosphate form) is a competitive inhibitor of viral DNA polymerase as above and leads to chain termination as it lacks 3'OH group	Genital HSV Primary Recurrent Suppression Encephalitis VZV–CMV prophylaxis	400mg p.o. tid × 7-10d (or) 5mg/kg i.v. q8h 400mg p.o. tid × 5d 400mg p.o. bid 10mg/kg i.v. q8h x 14d 800mg p.o. 5id × 7d (or) 10mg/kg i.v. q8h × 7d 800mg p.o. qid–5id
Valacyclovir	L-valyl ester of acyclovir converted to parent form prior to phosphorylation	Similar to acyclovir but with enhanced bioavailability of parent compound	Recurrent genital herpes Herpes–Zoster in immuno-compromised	500mg p.o. bid × 5d 1g p.o. tid × 7d

Famciclovir and Penciclovir	Acyclic analogue of guanosine	Penciclovir tri-phosphate competitively inhibits viral DNA polymerase	Herpes–zoster Recurrent genital HSV	500mg p.o. tid × 7d 125mg p.o. bid × 5d
Ganciclovir	Acyclic analogue of guanosine	Phosphorylation to tri-phosphate form competitively inhibits viral DNA polymerase	CMV retinitis CMV pneumonia transplant recipients	5mg/kg i.v. q12h × 14–21d followed by maintenance therapy (5mg/kg i.v. qd) 5mg/kg i.v. q12h × 10–14d with i.v. immunoglobulins
Foscarnet	Pyrophosphate (ppi) analogue	Blocks ppi binding site on the viral DNA polymerase—interferes with ppi exchange from dNTPs and inhibit noncompetitively	CMV retinitis Acyclovir resistant HSV or VZV Invasive HHV-6 disease	60mg/kg i.v. q8h (or) 90mg/kg i.v. q12h × 14–21d 40mg/kg i.v. q8h (or) 60mg/kg i.v. q12h × 14–21d 60mg/kg i.v. q8h × 14–21d
Cidofovir	Acyclic phosphonate nucleotide analogue	Exist in mono phosphorylated form, diphosphate form is a competitive inhibitor of viral DNA polymerase	CMV retinitis	5mg/kg i.v. qw × 2weeks, then every other week

(Contd.)

Table 5.1   Contd.

Antiviral agent	Property	Mechanism of action	Indication	Dosage
Sorivudine	Nucleoside analogue (used for VZV infection)	Sorivudine triphosphate is a competitive inhibitor of viral DNA polymerase, but do not act as chain terminator	Herpes zoster in immunosupp-ressed host	40mg p.o. qd × 7d
Trifluridine	Halogenated nucleoside analogue	Triphosphorylated form is a competitive inhibitor of viral DNA polymerase	HSV keratitis	1 drop of 1% solution q2h (max. 9 drops/d) × 10d
Idoxuridine	Pyrimidine nucleoside analogue	Anabolically phosphorylated to triphosphate form is a competitive inhibitor of viral DNA polymerase	HSV keratitis	1 drop q4h
Anti- retroviral agents				
Nucleoside analogue reverse transcriptase inhibitors (NRTIs)				
Zidovudine (AZT)	Azide group substituting at 3′ on ribose ring	Triphosphorylated form acts as competitive inhibitor of HIV reverse transcriptase (RT), by chain termination	HIV-1 and - 2 infection Mother-foetus transmission prophylaxis	200mg p.o. tid (or) 300mg p.o. bid Mother: 500-600mg/d after 14week of pregnancy, hen 2mg/kg i.v. × 1h

				followed by 1mg/kg/h i.v. during labour pain Infant: 2 mg/kg p.o. q6h × 6week
Didanosine (ddI)	Purine nucleoside analogue	Phosphorylated and converted to ddATP which is a competitive inhibitor of HIV-RT, by chain termination	HIV-1 and -2 infection	Tab: >60kg, 200 mg p.o. bid <60 kg, 125 mg p.o. bid
Zalcitabine (ddC)	Pyrimidine analogue	Phosphorylation by cell cycle independent enzyme active in both T cells and dendritic cells	HIV-1 and -2 infection	0.75 mg p.o. tid
Stavudine (d4T)	Thymidine analogue	Similar to AZT except monophosphorylation that is not a limiting factor by thymidine kinase	HIV-1 and -2 infection	> 60 kg, 40 mg p.o. bid < 60 kg, 30 mg p.o. bid
Lamivudine (3TC)	(-) enantiomer of a cytidine analog with sulphur at 3' in furanose ring	Lamuvidine triphosphate competitively inhibits HIV-RT	HIV-1 and -2 infection	150 mg p.o. bid

(Contd.)

Table 5.1   Contd.

Antiviral agent	Property	Mechanism of action	Indication	Dosage
Non-nucleoside analogue reverse transcriptase inhibitors (NNRTIs)				
Nevirapine	Muscarinic receptor antagonist	Noncompetitively inhibit by allosteric binding of the viral RT	HIV-1 infection	200 mg p.o. qd× 2 weeks, then 200 mg p.o. bid
Delavirdine	Bis-piperazine derivative (BHAP)	Noncompetitively inhibit by allosteric binding of the viral RT	HIV-1 infection	400 mg p.o. tid
HIV Protease inhibitors (PIs)				
Saquinavir	Hydroxy ethylamine derived peptidomimetic inhibitor	Bind to HIV-1 dimeric aspartyl protease that cleaves *gag* and *gag-pol* polypeptide precursors – inhibit virus replication	HIV-1 and –2 infection	600 mg p.o. tid
Indinavir	Peptidomimetic inhibitor	Similar to Saquinavir	HIV-1 and –2 infection	800 mg p.o. tid
Ritonavir	C2 symmetric HIV-PI	Binds C2 symmetric active site of HIV protease homodimer	HIV-1 and –2	600 mg p.o. bid

| Nelfinavir | Butyl carboxamide methane sulfonic acid salt | | HIV-1 and –2 infection | 750 mg p.c. tid (or) 1250 mg bid |
| Amprenavir (formerly V×478) | Hydroxyl ethyla-mine sulfonamide peptidomimetic | | HIV-1 and –2 | 1200 mg p.o. bid |

Other Agents

Amantadine	Tricyclic amine	Interferes with viral ion channel protein M2 coded by gene segment 7, and inhibits viral uncoating	Influenza A virus Prophylaxis Treatment	100 mg p.c. bid during risk period 100 mg p.c. bid × 5d
Rimantadine	Tricyclic amine	Interferes with viral ion channel protein M2 coded by gene segment 7, and inhibits viral uncoating	Influenza A virus Prophylaxis Treatment	200 mg p.c. qd during risk period 200 mg p.c. qd × 5d
Oseltamivir and Zanamavir	Neuraminidase inhibitor	Inhibit release of virus from host cell surface and spread of virus	Influenza virus A and B	Experimental

Investigational PIs include Tipranavir (dihydropyrone class non-peptide HIV PI), Atazanavir (azapeptide aspartyl PI) HIV-1 entry inhibitor – Pentafuside (T-20) interferes with HIV gp41 mediated cell fusion of virus

(Contd.)

Table 5.1 Contd.

Antiviral agent	Property	Mechanism of action	Indication	Dosage
Ribavirin	Synthetic guanosine analog	Complex and multifactorial, triphosphorylated interferes with capping and elongation of mRNA—inhibit viral RNA polymerase	Respiratory syncytial virus (RSV)	Aerosol (20 mg/ml) at 12-18h/d × 3-7d
Pleconaril	Experimental broad spectrum agent		Picornavirus, rhinovirus, coxsackievirus	400 mg p.o. tid × 7d
			Enterovirus	200-400 mg p.o. tid × 7d
Interferon-α	Human cytokine	Interferon- α and β are produced by all cells, and γ by T-cells and NK cells–induces antiviral state by binding to receptors on cell surface and inhibit viral replication	Chronic hepatitis B virus (HBV)	5 mU/d (or) 10 mU tiw s.c. or i.m. × 4 mon
			HCV	3 mU s.c. or i.m. tiw × 6 mon
Human immunoglobin		Hepatitis A post- exposure prophylaxis		0.02 ml/kg i.m. within 14d of exposure
		Measles post-exposure prophylaxis		0.25 ml/kg i.m. within 6d of exposure

CMV immunoglobin	Prophlaxis of CMV disease in transplantation recipients	150 mg/kg i.v. within 72 h of transplantation then 2, 4, 6, 8, 12 and 16 week post transplantation
Hepatitis B immunoglobin	Prophylaxis of blood exposure Infants of HBsAg positive mothers	0.06 ml/kg i.m. + vaccine 0.5 ml i.m. + vaccine
VZV immunoglobin	Post-exposure prophylaxis	125 U/10kg i.m. up to 625 U once
Rabies immunoglobin	Post-exposure prophylaxis	20 IU/kg i.m. + vaccine

q-every; qd- every day; bid-twice daily; d-day; p.o.-oral administration; tid-three times daily; i.m.-intramuscular; i.v.-intravenous; s.c.-subcutaneous; h-hour

It also has antiviral activity against other herpesviruses such as VZV, CMV and EBV, although the mechanism is not so well understood in these cases. Foscarnet is the preferred drug in the treatment of acyclovir-resistant strains.

Valacyclovir

Valacylovir is an ester of acyclovir that is well absorbed. Its bioavailability is 2–5 times greater than acyclovir. It is used for the treatment and suppression of genital herpes infection.

Famciclovir

Famciclovir is the prodrug of penciclovir which is the active form and a guanosine analogue. It has a very high bioavailability of 77%. It is converted into penciclovir by a two-step process. The first step occurs in the gut and the second step in the liver. It has a long half-life in the gut. It has a higher affinity for HSV thymidine kinase than acyclovir but a lower affinity for HSV DNA polymerase than acyclovir. It acts as an inhibitor of viral DNA polymerase and also as a chain terminator. At present famciclovir is licensed for the treatment of shingles and the dosage is 250 mg tds. It is also used for the treatment and suppression of genital herpes infection.

Ganciclovir

Ganciclovir is a guanine nucleoside chemically related to acyclovir (figure 5.1, Table 5.1). It acts as a chain terminator in subsequent termination of viral DNA replication. The active form is thought to be the triphosphate. CMV does not specify TK and the initial phosphorylation of ganciclovir is thought to be mediated by other cellular enzymes. Ganciclovir has potent in vitro activity against all herpesviruses, including CMV. It has some activity against other DNA viruses such as vaccinia and adenovirus. Ganciclovir is more active against CMV than acyclovir. Ganciclovir has been shown to be of value in treating severe CMV infections in the immunocompromised, especially in conjunction with hyperimmune immunoglobulin. Reversible neutropenia is the most frequent adverse reaction. Ganciclovir resistance has been reported in immunocompromised patients being treated for CMV disease and is thought to be due to lack of phosphorylation of the drug by CMV-infected cells. A recent prospective study estimated that 8% of patients receiving ganciclovir for more than 3 months developed resistant CMV.

Ribavirin

Ribavirin is a synthetic triazole nucleoside and the active form is ribavirin triphosphate. It is not incorporated into the primary structure of DNA or RNA during cellular synthesis of nucleic acids. In the case of influenza viruses, it inhibits the 5´ capping of viral mRNAs. It has also been shown to inhibit influenza viral RNA polymerase complex. It has further been postulated that ribavirin triphosphate inhibits several steps in viral replication and this phenomenon may explain the failure to detect viral isolates that are resistant to ribavirin. Ribavirin has been shown to possess activity against both DNA and RNA viruses in infected cells. It has been found to have activity against adenoviruses, herpesviruses, CMV, vaccinia, influenza A and B, parainfluenza 1, 2, 3, measles, mumps, RSV and rhinovirus. Ribavirin has made a major contribution to the therapy of children infected with RSV, since it is given as an aerosol in hospitals. It has also been shown to be effective against influenza A and B. It has also been reported to be of value in the treatment of Lassa fever, hantavirus disease and hepatitis C.

Zidovudine (AZT)

AZT is a synthetic analogue of thymidine. It requires conversion to the triphosphate form by cellular enzymes. It inhibits viral reverse transcriptase by acting as a chain terminator. Viral reverse transcriptase is 100 times more susceptible to inhibition by zidovudine triphosphate than host cellular DNA polymerase. Once incorporated into the viral DNA chain, viral DNA synthesis is terminated, as no more phosphodiester bonds can be formed. AZT is active in vitro against many human retroviruses, including HTLV-I and HIV. AZT is currently indicated for the management of patients with HIV infection who have impaired immunity (T4 cell count of 400–500 or less). AZT has clearly been shown to prolong the life of individuals infected with HIV. It has also been shown to be of benefit for the treatment of symptomless individuals although being controversial.

Lamivudine

Lamivudine is a potent reverse transcriptase inhibitor. It is generally well tolerated by patients. Nowadays, it forms an essential component in the combination therapy of HIV patients. It had also been approved for the treatment of chronic hepatitis B recently.

Foscarnet

Foscarnet is a pyrophosphate analogue and unlike nucleoside analogues, foscarnet does not need to be activated by cellular or viral kinases. Foscarnet binds directly to the pyrophosphate-binding sites of RNA or DNA polymerases. Foscarnet is difficult to use as it must be given continuously intravenously via an infusion pump. It is used for the treatment of CMV retinitis in AIDS patients receiving AZT therapy, as it does not have overlapping toxicity with AZT. It is also used in the treatment of AZT resistant HSV infections. Its major adverse effect is on renal function.

Amantidine

Amantidine inhibits the growth of influenza viruses in cell culture and in experimental animals. It is only effective against influenza A, and some naturally occurring strains of influenza A are resistant to it. The mechanism of action of amantadine is not known. It is thought to act at the level of virus uncoating. The compound has been shown to have both therapeutic and prophylactic effects. Amantidine significantly reduces the duration of fever (51 hours as opposed to 74 hours) and illness. The compound also conferred 70% protection against influenza

A when given prophylactically. Amantidine can occasionally induce mild neurological symptoms such as insomnia, loss of concentration and mental disorientation. However, these symptoms quickly developed in susceptible individuals and cease when treatment is stopped. The therapeutic and prophylactic activity of amantidine is now generally accepted and numerous analogues of this compound have been prepared. Rimantadine is not as effective as amantadine but is less toxic. One factor that limits the usefulness of amantidine and rimantidine is the rapid development of resistance of these molecules in 30% of patients. These resistant mutants have been reported to be as capable of being transmitted and causing disease as the wild virus.

Zanamivir

The rational approach to drug design has led to the design of several potent inhibitors of influenza neuraminidase of which zanamivir and oseltamivir are licensed for the treatment of influenza A and B infections. In clinical trials, both agents have demonstrated efficacy with minimal side effects. Because of its poor bioavailability, zanamivir must be given by inhalation while oseltamivir can be given orally. Because selection of drug-resistant mutants characterized by changes in NA requires prolonged passage in tissue culture, development of zanamivir-resistant viruses is not expected to occur readily in patients. The available information suggests that mutants may be less stable in vivo. The significance of changes in hemagglutinin remains to be evaluated. Overall, the NA family of anti-influenza drugs is showing considerable promise; resistant variants do not occur readily and may be biological cripples.

Immunoglobulins

Immunoglobulins are available in three different formulations; intramuscular form, IVIG and hyperimmune globulins against individual viruses. Immunoglobulins are more effective when used prophylactically rather than therapeutically. Currently, HNIG is used primarily for the prevention of hepatitis A. HNIG can also be given to non-immunized contact of measles. Hyperimmune globulins are used for the post-exposure prevention of hepatitis B, chicken pox and rabies. It has also been used in the treatment of arenavirus infections, Crimean-Congo haemorrhagic fever and Rift valley fever. CMV Ig is given prophylactically to seronegative recipients of kidneys from seropositive donors. The use of prophylactic CMV Ig in BMT patients is controversial. CMV IVIg is used in conjunction with ganciclovir in the treatment of CMV pneumonitis. IVIg is also used in the treatment of chronic enteroviral meningoencephalitis in children with agammaglobinaemia.

Anti-retroviral Agents

The main groups of substances described are:

    a. *Nucleoside analogue reverse transcriptase inhibitors* (e.g. AZT, ddC, ddI and lamuvidine)

    b. *Non-nucleoside analogue reverse transcriptase inhibitors* (e.g. Nevirapine)

    c. *HIV protease inhibitors* (e.g. Ritonavir and Indivavir are the most potent inhibitors of HIV replication to date)

Zidovudine (AZT) was the first antiviral agent shown to have beneficial effect against HIV infection (figure 5.1, Table 5.1). However, after prolonged use, AZT-resistant strains rapidly appear limiting the effect of AZT. Recent clinical trials reported significant benefit in the use of combination therapy over the use of monotherapy. The rationale for this approach is that by combining drugs that are synergistic, non-cross-resistant and no overlapping toxicity, may possibly reduce toxicity, improve efficacy and prevent resistance from arising. In fact, significant success has now been reported for trials involving multiple agents including protease inhibitors. The aim of anti-HIV therapy has now shifted from simply delaying the progression of disease to finding a permanent cure. We have now entered the era of highly active anti-retroviral therapy (HAART).

The current consensus is that one should give a potent combination of HAART agents right from the start when treatment is indicated. The most popular combination is AZT and lamivudine plus a protease inhibitor. Lamivudine has greater anti-retroviral activity than AZT and is active against many AZT-resistant strains without significant increase in toxicity. Among protease inhibitors, indinavir (IDV) is more potent than saquinavir and appears to have fewer drug interactions and short-term adverse effects than ritonavir. In currently recommended doses, AZT prophylaxis is well tolerated with health workers; short-term toxicity associated with higher doses primarily includes GI symptoms, fatigue and headache. In HIV-infected adults, 3TC can cause GI symptoms, and rarely pancreatitis. IDV toxicity includes GI symptoms and after prolonged use, mild hyperbilirubinaemia (10%), and kidney stones (4%).

It is generally agreed that treatment should be started when CD4 < 500 ml or viral load >5000–10000 copies/ml (bDNA assay). If CD4 count is >500/ml but viral load is >5000–10000 copies/ml (bDNA assay), then recommendations vary. It may be then advisable to treat those who are compliant and committed. The actual recommendation as to when to commence treatment, and the regimen to use, varies greatly between different countries. The details of indications and dosage regimens of various antiviral agents are depicted in Table 5.1.

ANTI-HIV THERAPY

The huge resources that had gone into HIV research had resulted in the development of a large number of anti-HIV agents. Therapy of HIV is complicated by the fact that the HIV genome is incorporated into the host cell genome and can remain there in a dormant state for prolonged periods until it is reactivated. Effective therapy must be directed against both free virus and virus-infected cells. Although a number of substances with in vitro anti-HIV activity have been described, only a few drugs exhibit anti-HIV activity in vivo at tolerable toxicities.

Monitoring Anti-HIV Therapy

Viral Load

Initiation   Viral load is now the preferred method of monitoring therapy. There should be <= 1 log reduction in viral load, preferably to less than 10,000 copies/ml HIV-RNA within 2–4 weeks after the commencement of treatment. If <0.5 log reduction in viral, or HIV RNA stays above 100,000, then the treatment should be adjusted by either adding or switching drugs.

Monitoring   Viral load measurement should be repeated every 4–6 months if the patient is clinically stable. If the viral load returns to 0.3–0.5 log of pre-treatment levels, then the therapy is no longer working and should be changed.

CD4 Count

Initiation   Within 2–4 weeks of starting the treatment, CD4 count should be increased by at least 30 cells/mm^3. If this is not achieved, then the therapy should be changed.

Monitoring   CD4 counts should be obtained every 3–6 months during periods of clinical stability, and more frequently should symptomatic disease occurs. If CD4 count drops to baseline (or below 50% of increase from pre-treatment), then the therapy should be changed.

INTERFERONS (IFNs)

There are three classes of interferons: alpha, beta and gamma. Interferon-α exists as at least 15 subtypes, the genes for which show 85% homology. IFNb1 shows 30% homology with IFNa. IFNb2 is now known as IL-6 and shows no homology with alpha or β1 types. IFN gamma is also a lymphokine and shows no homology with other types. IFNs mediate their action through specific receptors at hormone-like concentrations. Interferon-inducible response elements in the cellular genome are activated. There are two main types of IFN receptors, one for alpha and beta and the other for gamma.

IFNs are released from many cell types in response to virus infection, dsRNA, endotoxin, mitogenic and antigenic stimuli. Double-stranded RNA appears to be a particularly important inducer. Usually, good IFN inducers are viruses that multiply slowly and do not block the synthesis of host protein early or markedly damage the cells. IFN is usually assayed by determining its effect on the multiplication of a test virus, usually vesicular stomatitis virus, a rhabdovirus. Viral strains capable of high IFN production give rise to autointerference in endpoint assays. In general IFN gamma differs from the others in one aspect that it is released as a lymphokine from activated T-cells, natural killer (NK) cells and occasionally from macrophages.

Mechanism of Action

The antiviral effects of IFNs are exerted through several pathways:

1. Increased expression of Class I and Class II MHC glycoproteins, thereby facilitating the recognition of viral antigens by the immune system.
2. Immunoregulatory effects—activation of cells with the ability to destroy virus-infected targets; these include NK cells and macrophages. IFNs appear to drive a shift from humoral to cellular immunity.
3. Direct inhibition of viral replication.

Interferons are now being used for treatment of chronic HBV and HCV carriers who are at risk of progression to cirrhosis and HCC. The recommended regimen is stated in the table (Table 5.1). For those who relapse after treatment, they may be offered a second course and then put on maintenance therapy for 6–12 months. There is data to suggest that combination therapy with interferon and ribavirin is more effective than interferon for HCV infection alone.

VACCINES

Immunity in the case of viruses is based on the development of an immune response to specific antigens on the surface of the viral particles. Vaccines are available for the prevention of several significant human diseases. They may be made from live attenuated (virulence is radically diminished) or killed virus. Attenuated vaccines have an advantage in that they simulate a natural infection and stimulate longer-lasting antibody production, induce a good cell-mediated response and induce resistance at the point of entry. Killed viruses are made by purifying viral harvests before inactivation by some method that only minimally damages the viral proteins.

Requirements of a Good Vaccine

To be effective a vaccine should be capable of eliciting the following:

- Activation of antigen-presenting pells to initiate antigen processing and producing interleukins.
- Activation of both T and B cells to give a high yield of memory cells.
- Generation of Th and Tc cells to several epitopes, to overcome the variation in the immune response in the population due to MHC polymorphism.
- Persistence of antigen, probably on dendritic follicular cells in lymphoid tissue, where B memory cells are recruited to form antibody-secreting cells that will continue to produce antibody.

Two ingredients are needed for a successful immunization programme:

- A safe and effective vaccine
- An appropriate strategy with adequate vaccine coverage

The strategy required depends on whether the aim of the program is eradication, elimination or containment. *Eradication* is the complete extinction of the organism in question. In *elimination*, the disease disappears but the organism remains. *Containment* is the control of the disease to the point at which it no longer constitutes a public health problem.

When eradication or elimination is the aim, mass immunization in early life of both sexes is usually necessary. When containment is the aim, selective immunization of those most at risk is normally sufficient. The usual indications for selective immunization are travel, occupational risk, outbreak control and for individuals at special risk of severe illness. Herd immunity plays little part, as the virus continues to circulate widely among the unvaccinated.

Vaccination policy varies between different countries and Maternal antibodies present up to 6 months after birth, may interfere with the induction of an effective immune response against the vaccine in the infant. This should be duly taken into account when formulating a vaccination policy.

Different Types of Vaccines

Whole virus vaccines either live or killed, constitute the vast majority of vaccines in use at present. However, recent advances in molecular biology have provided alternative methods for producing vaccines. Listed below are the possibilities:

1. Live whole virus vaccines
2. Killed whole virus vaccines
3. Subunit vaccines; purified or recombinant viral antigen
4. Recombinant virus vaccines
5. Anti-idiotype antibodies
6. DNA vaccines

Live vaccines Live virus vaccines are prepared from attenuated strains that are almost or completely devoid of pathogenicity but are capable of inducing a protective immune response. They multiply in the human host and provide continuous antigenic stimulation over a period of time. Primary vaccine failures are uncommon and are usually the result of inadequate storage or administration. Another possibility is interference by related viruses as is suspected in the case of oral polio vaccine in developing countries. Several methods have been used to attenuate viruses for vaccine production.

- Use of a related virus from another animal—the earliest example was the use of cowpox to prevent smallpox.

- Administration of pathogenic or partially attenuated virus by an unnatural route—the virulence of the virus is often reduced when administered by an unnatural route. This principle is used in the immunization of military recruits against adult respiratory distress syndrome using enterically coated live adenovirus type 4, 7 and 21.

- Passages of the virus in an "unnatural host" or host cell—the major vaccines used in man and animals have all been derived this way. After repeated passages, the virus is administered to the natural host. The initial passages are made in healthy animals or in primary cell cultures. There are several examples of this approach: the 17D strain of yellow fever was developed by passage in mice and then in chick embryos. Polioviruses were passaged in monkey kidney cells and measles in chick embryo fibroblasts. Human diploid cells such as the WI-38 and MRC-5 are now widely used. The molecular basis for host range mutation is now beginning to be understood.

Inactivated whole virus vaccines   These were the easiest preparations to use. The preparation was simply inactivated. The outer virion coat should be left intact but the replicative function should be destroyed. To be effective, non-replicating virus vaccines must contain much more antigens than live vaccines that are able to replicate in the host. Preparation of killed vaccines may take the route of heat or chemicals. The chemicals used include formaldehyde or beta- propiolactone. The traditional agent for inactivation of the virus is formalin. Excessive treatment can destroy immunogenicity whereas insufficient treatment can leave infectious virus capable of causing disease. Another problem was that SV40 was occasionally found as a contaminant and there were fears of the potential oncogenic nature of the virus.

Potential Safety Problems

Live vaccines   Underattenuation ,Mutation leading to reversion to virulence, preparation instability, contaminating viruses in cultured cells and heat lability, pose potential safety problems, should not be given to immunocompromised or pregnant patients.

Killed vaccines   Incomplete inactivation and increased risk of allergic reactions due to large amounts of antigen involved, Present problems with vaccine development include failure to grow large amounts of organisms in laboratory and crude antigen preparations which often give poor protection.

Subunit Vaccines

Originally, non-replicating vaccines were derived from crude preparations of virus from animal tissues. As the technology for growing viruses to high titers in cell cultures advanced, it became practicable to purify virus and viral antigens. It is now possible to identify the peptide sites encompassing the major antigenic sites of viral antigens, from which highly purified subunit vaccines can be produced. Increasing purification may lead to loss of immunogenicity, and this may necessitate coupling to an immunogenic carrier protein or adjuvant, such as an aluminium salt. Examples of purified subunit vaccines include the HA vaccines for influenza A and B, and HBsAg derived from the plasma of carriers.

Recombinant Viral Proteins

Virus proteins have been expressed in bacteria, yeast, mammalian cells, and viruses. Recombinant hepatitis B vaccine is the only recombinant vaccine licensed at present.

An alternative application of recombinant DNA technology is the production of hybrid virus vaccines. The best known example is vaccinia; the DNA sequence coding for the foreign gene is inserted into the plasmid vector along with a vaccinia virus promoter and vaccinia thymidine kinase sequences. The resultant recombination vector is then introduced into cells infected with vaccinia virus to generate a virus that expresses the foreign gene. The recombinant virus vaccine can then multiply in infected cells and produce the antigens of a wide range of viruses. The genes of several viruses can be inserted, so the potential exists for producing polyvalent live vaccines. HBsAg, rabies, HSV and other viruses have been expressed in vaccinia.

Hybrid virus vaccines are stable and stimulate both cellular and humoral immunity. They are relatively cheap and simple to produce. Being live vaccines, smaller quantities are required for immunization. As yet, there are no accepted laboratory markers of attenuation or virulence of vaccinia virus for man. Alterations in the genome of vaccinia virus during the selection of recombinant may alter the virulence of the virus. The use of vaccinia also carries the risk of adverse reactions associated with the vaccine and the virus may spread to susceptible contacts. At present, efforts are being made to attenuate vaccinia virus further and the possibility of using other recombinant vectors such as attenuated poliovirus and adenovirus is being explored.

Synthetic Peptides

The development of synthetic peptides that might be useful as vaccines depends on the identification of immunogenic sites. Several methods have been used. The best known example is foot-and-mouth disease, where protection was achieved by immunizing animals with a linear sequence of 20 amino acids. Synthetic peptide vaccines would have many advantages. Their antigens are precisely defined and free from unnecessary components which may be associated with side effects. They are stable and relatively cheap to manufacture. Furthermore, less quality assurance is required. Changes due to natural variation of the virus can be readily accommodated, which would be a great advantage for unstable viruses such as influenza.

Synthetic peptides do not readily stimulate T cells. It was generally assumed that, because of their small size, peptides would behave like haptens and would therefore require coupling to a protein carrier which is recognized by T-cells. It is now known that synthetic peptides can be highly immunogenic in their free form provided they contain, in addition to the B cell epitope, T- cell epitopes recognized by T-helper cells. Such T-cell epitopes can be provided by carrier protein molecules, foreign antigens or within the synthetic peptide molecule itself.

Anti-idiotype Antibodies

The ability of anti-idiotype antibodies to mimic foreign antigens has led to their development as vaccines to induce immunity against viruses, bacteria and protozoa in experimental animals. Anti-idiotypes have many potential uses as viral vaccines, particularly when the antigen is difficult to grow or is hazardous. They have been used to induce immunity against a wide range of viruses, including HBV, rabies, Newcastle disease virus and FeLV, reoviruses and polioviruses.

DNA Vaccines

Recently, encouraging results were reported for DNA vaccines whereby DNA coding for the foreign antigen is directly injected into the animal so that the foreign antigen is directly produced by the host cells. The technique that is being tested in humans involves the direct injection of plasmids—loops of DNA that contain genes for proteins produced by the organism being targeted for immunity. Once injected into the host's muscle tissue, the DNA is taken up by host cells, which then start expressing the foreign protein. The protein serves as an antigen that stimulates an immune response and protective immunological memory.

Important concerns over DNA vaccines are:

1. Delivery of the DNA to cells is still not optimal, particularly in larger animals.
2. Possibility of the vaccine's DNA to be integrated into host chromosomes will turn on oncogenes or turn off tumour suppressor genes.
3. Extended immunostimulation by the foreign antigen could in theory provoke chronic inflammation or autoantibody production.

Presentation of Immunogenic Proteins and Peptides

Proteins separated from virus particles are generally much less immunogenic than the intact particles. This difference in activity is usually attributed to the change in configuration of a protein when it is released from the structural requirements of the virus particle. Many attempts have been made to enhance the immunogenic activity of separated proteins.

Adjuvants   They are used to potentiate the immune response and perform two functions

- To localize and slowly release antigen at or near the site of administration.

- To activate antigen presenting cells (APCs) to achieve effective antigen processing or presentation

Materials that have been used include aluminium salts, Mineral oils , Mycobacterial products, (e.g. Freud's adjuvants).

In the most successful procedure, a mixture of the plant glycoside saponin, cholesterol and phosphatidylcholine provides a vehicle for presentation of several copies of the protein on a cage–like structure. Such a multimeric presentation mimics the natural situation of antigens on microorganisms. These immunostimulating complexes have activities equivalent to those of the virus particles from which the proteins are derived, thus holding out great promise for the presentation of genetically engineered proteins.

Table 5.2 Immunization of viral diseases—vaccines, types and routes of administration.

Vaccine	Type	Cell substrate	Route of administration	Comment
Adenovirus	Live	Human diploid fibroblasts (WI-38)	Oral	Enteric capsule types 4 & 7
Hepatitis A	Inactivated	Human diploid fibroblasts (MRC-5)	Intramuscular	Intended for high-risk individuals or in endemic settings
Hepatitis B	Subunit (HBsAg)	Yeast (rDNA)	Intramuscular	Effective in reduction of HBV and hepatocellular carcinoma
Influenza A&B	Inactivated	Embryonated hen eggs	Intramuscular	Whole or split virion, annual vaccine required
Japanese encephalitis	Inactivated	Mouse brain	Subcutaneous	Allergic reaction with delayed onset can occur
Measles	Live attenuated	Chicken embryo fibroblasts	Subcutaneous	Initial dose given at 1year of age
Mumps	Live attenuated	Embryonated hen eggs and chicken eggs	Subcutaneous	In combination with measles and rubella
IPV	Inactivated	Vero monkey kidney cells	Subcutaneous	Enhanced potency formulation
OPV	Live	Monkey kidney cells	Oral	Supplanted in US by all IFV schedule

(Contd.)

Table 5.2   Contd.

Vaccine	Type	Cell substrate	Route of administration	Comment
Rabies	Inactivated	Human diploid fibroblasts (MRC-5)	Intramuscular or intradermal	Given with rabies immunoglobulin in post-exposure prophylaxis
Rubella	live	Human diploid fibroblasts (WI-38)	Subcutaneous	In combination with measles and mumps vaccines
Varicella-zoster	live	Human diploid fibroblasts (MRC-5)	Subcutaneous	One dose for children <13 years, 2 doses for >13 years old
Yellow fever virus	Live	Embryonated hen eggs	Subcutaneous	Booster doze every 10 years

HIV vaccine candidates in clinical trial:

Gp120 – subtypes B/B, B/E—Phase III trial in US and Thailand

ALVAC-HIV – B, E—Phase II trials in US, Haiti, Brazil

ALVAC-HIV – A—Phase I trial in Uganda

Lipopeptides LP5, LP6 – subtype B – Phase I trial in France

Vaccinia TBC-3B, subtype B—Phase I trial in US

DNA-HIV – subtype B—Phase I trial completed

MVA-HIV – subtype A—Phase I trial in UK and Kenya

DNA – HIV, Adenovirus-HIV, Subtype B—Phase I trial in US

ALVAC-HIV, recombinant canarypox expressing multiple HIV genes; MVA-HIV—modified vacinia Ankara, an attenuated vaccinia vector, expressing multiple HIV genes; TBC–3B— attenuated vaccinia vector expressing multiple HIV genes.

Similar considerations apply to the presentation of peptides. It has been shown that by building the peptide into a framework of lysine residues so that eight copies instead of one copy are present, the immune response induced was of a much greater magnitude. A novel approach involves the presentation of the peptide in a polymeric form combined with T cell epitopes. The sequence coding for the foot-and-mouth disease virus peptide was expressed as part of a fusion protein with the gene coding for the Hepatitis B core protein. The hybrid protein, which forms spherical particles 22 nm in diameter, elicited levels of neutralizing antibodies against foot-and-mouth disease virus that were at least a hundred times greater than those produced by the monomeric peptide. The availability of viral vaccines, its types, source and route of administration are stated in Table 5.2.

REVIEW QUESTIONS

Write short notes on

    a. AZT

    b. Nucleoside analogues

    c. Acyclovir

    d. Foscarnet

    e. Amantidine

    f. Interferons

    g. Reverse transcriptase

    h. CD4 count/CD8 count

    i. HAART

    j. Vaccine

    k. IPV and OPV

    l. DNA vaccines

    m. MMR vaccine

    n. Subunit vaccine

o.   Influenza virus vaccine

p.   Treatment of herpesvirus infections

q.   Hepatitis B virus vaccine

r.   Rabies vaccine

GLOSSARY

Acquired Immunodeficiency Syndrome (AIDS) Severe manifestation of infection with the human immunodeficiency virus (HIV). Predisposes patients to life-threatening infections by numerous opportunistic pathogens and neoplasms, which in the presence of HIV infection, constitute an AIDS diagnosis. In addition, a CD4 count below $200/mm^3$ in the presence of HIV infection constitutes an AIDS diagnosis.

Acromegaly A chronic disease caused by hypersecretion of the pituitary growth hormones. It is characterized by enlargement of many parts of the skeleton, especially the nose, ears, jaws, fingers and toes.

Acute Short-term, intense symptomatology or pathology, as distinct from chronic. Many diseases have an acute phase (like HIV seroconversion disease) and a chronic phase. This distinction is sometimes used in treatments, as in high-dose ganciclovir for acute (or induction) treatment of CMV retinitis, followed by a lower dose for chronic (or maintenance) treatment.

Acyclovir (Zovirax) An antiviral drug used in the treatment of herpes simplex virus 1 (HSV-1, fever blisters, cold sores), herpes simplex virus 2 (HSV-2, genital herpes) and herpes zoster (shingles), and sometimes for acute varicella-zoster virus (chickenpox); its inhibitory action results from its selective phosphorylation by the thymidine kinase and subsequent inhibition of the virus-encoded DNA polymerase.

Adenopathy Enlargement of glands, especially the lymph nodes.

Adverse reactions Any undesirable effect of a medication. All drugs may cause such reactions, so that periodic monitoring is necessary to detect any that do occur, even though their occurrence may be uncommon.

Alanine aminotransferase (Alt) A liver enzyme, measured through a blood test, that indicates the health of the liver. Lower counts are better. Levels may go up because of hepatitis and other infections, or because of drug toxicities.

AIDS Dementia Complex A degenerative neurological condition, with a wide variety of clinical presentations, including loss of coordination, mood swings and loss of inhibitions, and finally widespread cognitive deficit. AIDS dementia complex is generally thought to be caused by HIV itself.

AIDS Vaccine Evaluation Unit (AVEU) One of NIAID's clinical research sites for vaccine trials.

AIDS-Related Complex (ARC) 1) A term, not officially defined or recognized by the Centres for Disease Control and Prevention, that has been used to describe a variety of symptoms and signs found in some persons infected with HIV. These may include recurrent fevers, unexplained weight loss, swollen lymph nodes, and/or fungus infection of the mouth and throat. Also commonly described as symptomatic HIV infection. 2) Symptoms that appear to be related to infection by the HIV virus. They include an unexplained, chronic deficiency of white blood cells (leukopenia) or a poorly functioning lymphatic system with swelling of the lymph nodes (lymphadenopathy) lasting for more than three months without the opportunistic infections required for a diagnosis of AIDS.

Amino acid   Any one of 20 or more organic acids, some of which are the building blocks for proteins and are necessary for metabolism and growth.

Anaemia   A condition in which there is a decreased volume of red cells in the blood. There are many causes for anaemia, including drug toxicities and chronic infections. The most common way in which anaemia is measured is by the titre of haemoglobin (Hgb) in peripheral venous blood.

Anergic   Refers to the state of being so immunologically suppressed that one is unable to produce cutaneous delayed type hypersensitivity reaction (DTH). Such patients will usually not test positive for TB on a PPD (Mantoux) test.

Anorexia   Lack of or complete loss of appetite for food; aversion to food.

Antenatal   Before the time of birth.

Antibiotic   A chemical substance that kills or inhibits the growth of bacteria; some antibiotics are used to treat infectious diseases.

Antibody   A protein molecule in the blood serum or other body fluids that destroys or neutralizes bacteria, viruses, or other harmful toxins. Antibody production occurs in response to the presence of an antagonistic, usually foreign substance (antigen) in the body. They are members of a class of proteins known as immunoglobulins that are produced and secreted by B-lymphocytes in response to stimulation by an antigen. The antigen/antibody reaction forms the basis of humoral (non-cellular) immunity.

Antigen   Any substance that antagonizes or stimulates the immune system to produce antibodies, proteins that fight antigens. Antigens are often foreign substances such as bacteria or viruses that invade the body.

Antigen Presenting Cell (APC)   A white blood cell that devours foreign bodies, breaks them down, and carries characteristic antigen peptides to its surface. The foreign antigen, complexed with MHC I or II is presented to CD4 or CD8 to initiate an immune response specific to that peptide.

Antigenemia   The presence of detectable amounts of an antigen in the blood.

Antigenic Drift   The slow, sequential change in the antigenic nature of one or more viral surface proteins.

Antigenic Shift   A major change in the antigenic nature of two or more viral surface proteins to create a serologically distinct virus.

Antigenicity   The ability of an antigen to combine with antibodies and T-cell receptors to invoke a reaction from the immune system.

Antiretroviral   A substance, drug, or process that destroys a retrovirus, or suppresses its replication. Often used to describe a drug active against HIV.

Antiviral   A substance, drug, or process which destroys a virus or suppresses its replication. Can apply to anti-HIV activity, or other viruses, such as herpes or CMV.

Aphid   Phytophagous insect belonging to the family Aphidae and a common vector of many plant viruses.

Apoptosis   Cell suicide. Thought to be primarily a way that the body clears out immune cells that respond to the body's own proteins. Apoptosis involves a complete physical destruction of a cell, driven by enzymes. Apoptosis also occurs when one receptor on a CD4 cell is triggered without the normal "co-activation" signal. Abnormal apoptosis may be elevated in persons with HIV.

Arbovirus   Obsolete term used to describe any virus of vertebrates that is transmitted by an arthropod vector.

ARC   A term never officially defined by the CDC which has been used to describe a variety of symptoms and signs found in some persons infected with HIV. These may include a decrease in CD4 cells, recurrent fevers, unexplained weight loss, swollen lymph nodes, and/or fungus infection of the mouth and throat. Most of the clinical findings which were formerly denoted as ARC are now in groups 3 or 4 of the CDC AIDS classification system, although the term ARC is still popularly used to describe these symptoms and diagnoses. ARC is also commonly described as symptomatic HIV infection.

Aseptic   Without the presence of disease-causing microorganisms.

Aseptic meningitis   An inflammation of the meninges caused by a noncellular infectious agent (a virus).

Aspartate aminotransferase   A liver enzyme, measured through a blood test, that indicates the health of the liver. Lower counts are better. Levels may go up because of hepatitis and other infections, or because of drug toxicities.

Asymptomatic Without signs or symptoms of disease.

Asymptomatic Infection An infection or phase of infection, without symptoms.

Ataxia Problems with coordination or proper use of muscles.

Attenuated Virus A weakened virus with reduced ability to infect or produce disease. Some vaccines are based on attenuated viruses.

Autoimmune Diseases Diseases caused when an organism's own immune system attacks its own cells.

AZT Also called zidovudine, retrovir or azidothymidine. A thymidine (genetic building block) analogue that suppresses replication of HIV. It is the only FDA-approved drug for the initial treatment of HIV infection. Adverse side effects may include anaemia, leukopenia, muscle fatigue, muscle wasting, nausea and headaches.

Bacteriophages Viruses that infect bacteria.

Baseline characteristic A variable that is measured, observed, or assessed on a patient at or shortly before treatment assignment and the initiation of treatment.

Baseline data The set of data collected on a specific patient or set of patients prior to randomization.

Baseline examination An examination that is carried out as part of the baseline visit and that is designed to assess a patient's eligibility for enrollment into the trial and to produce required baseline data.

bDNA (Branched DNA Assay) A DNA test for detecting and measuring HIV in the blood plasma of people with HIV.

The bDNA test is faster and probably more accurate than plasma culture. The test currently used is less sensitive than PCR, another new test. bDNA testing may eventually be useful to monitor the effectiveness of anti-HIV drugs and to gauge HIV disease progression. It is not yet FDA-approved, nor is it widely available.

Beta carotene A form of carotene, precursor to vitamin A; a red-orange pigment found in plants and plant-eating animals, and also found in dark green and dark yellow fruits and vegetables. Beta carotene may have beneficial effects on the immune system.

B-Lymphocytes B-lymphocytes are blood cells of the immune system derived from the bone marrow and spleen involved in the production of antibodies. B-lymphocytes float through all body fluids, are able to detect the presence of foreign invaders, and produce antibodies on their own and when primed by T-lymphocytes. B-lymphocytes can later differentiate into plasma and memory cells. B-cells mediate the "humoral" immune response.

Bronchitis An inflammation of the bronchial tubes, generally accompanied by coughing, pain, or shortness of breath.

Burst Size The increase in the number of virions (infective centres) produced during a one-step growth cycle.

Candida albicans A yeast-like fungi, commonly found in the normal flora of the mouth, skin, intestinal tract, and vagina. Generally, candida is harmless, but can become clinically infectious in immunocompromised people.

Candida kruseii Another candida species, similar to Candida albicans, but often less susceptible to the common drugs used to treat *Candida albicans.*

Candidemia *Candida albicans* in the blood.

Candidiasis An infection with a fungus of the *Candida* family, generally *C. albicans*. The most common sites for candidiasis are the mouth, the throat and the vagina.

Capsid The protein covering of some viruses—made up of capsomeres that either surrounds the viral nucleic acid or complexes with it to form the nucleocapsid; may stimulate the body's immune response.

Capsomere Any of the proteins that comprise the capsid or the protein part of the nucleocapsid of a virion.

CD4 (T4) A protein embedded in the cell surface of helper T-lymphocytes; also found to a lesser degree on the surface of monocyte/macrophage, Langerhans cells, astrocytes, keratinocytes and glial cells. One of the ways HIV invades cells is by first attaching to the CD4 molecule (CD4 receptor).

CD4 Cell Helper T-cell, responsible for coordinating much of the immune response. CD4 cells are one of the main targets damaged by HIV.

CD4 Count The number of T-helper lymphocytes per cubic millimetre of blood. The CD4 count is a good predictor of immune health. A CD4 count less than 200 qualifies as a diagnosis of AIDS.

CD8 (T8) A protein embedded in the cell surface of killer and suppresser T-lymphocytes.

CD8 Count The number of killer/suppresser T-lymphocytes in a cubic millimetre of blood.

Cell Lines Specific cell types artificially maintained in the laboratory (in vitro) for scientific purposes.

Cell-Mediated Immunity (CMI) A branch of the immune system responsible for the reaction to foreign material by specific defence cells (T-lymphocytes, killer cells, macrophages and other white blood cells) rather than antibodies.

Central Nervous System (CNS) Composed of the brain, spinal cord, and its coverings (meninges).

Cerebrospinal Fluid (CSF) Fluid that bathes the brain and spinal cord.

Cervical dysplasia The development of abnormal tissue on the cervix, the lower part of the uterus; may progress to cancer of the uterus.

Cervix The cylindrical, lower part of the uterus leading to the vagina.

Chemoprophylaxis Prevention of disease by chemical means.

Chemotherapy The treatment of disease by chemical agents; usually, but not always refers to cancer treatment.

Chromosomes A condensed DNA structure normally found in the nucleus of a cell.

Chronic Referring to a process, such as a disease process, that occurs slowly and persists over a long period of time; opposite of acute.

Cloning vector A bacteriophage or plasmid that can be recombined with a foreign piece of DNA before it is replicated (cloned) in a host cell.

Cognitive Pertaining to thought, awareness, or the ability to rationally apprehend the world and abstract meaning.

Complementary DNA (cDNA) ssDNA molecule that is complementary to the base sequence of the ss-template from which it was transcribed; usually applied to DNA transcribed from RNA using reverse transcriptase.

Concatemer Long molecules containing a number of identical monomers linked in a head-to-tail fashion; such DNA or RNA molecules often occur as intermediates in virus replication.

Condyloma acuminata Clusters of warts that occur on the genitalia or perineum. Also called venereal warts.

Congenital rubella syndrome The developmental problems and the permanent defects that result from infection of a foetus by Rubella virus.

Conjunctivitis Inflammation of the protective membrane surrounding the eye.

Coryza Inflammation of the mucous membranes of nasal passages; also known as head cold.

Coxsackie viruses A group of enteroviruses belonging to the Picornaviridae family; they are named after Coxsackie, New York.

Cyanophages Phages isolated from blue-green algae.

Cytolytic Having a destructive effect on cells, causing them to erupt or burst open.

Cytomegalovirus (CMV) A herpesvirus which is a common cause of opportunistic diseases in people with AIDS and other people with immune suppression. While CMV can infect most organs of the body, people with AIDS are most susceptible to CMV retinitis and colitis.

Cytopatheic Effect   Localized or generalized degenerative changes or abnormalities of cells in culture or tissue, such as embryonated eggs or cell lines, caused by the activity of a virus.

Cytoskeleton   Protein filaments that extend through the cytoplasm of cells and enable them to move and change shape.

Cytotoxic   An agent or process which is toxic to cells that results in suppression of function or cell death.

Cytotoxic T Lymphocyte (CTL)   A lymphocyte that is able to kill foreign cells that have been marked for destruction by the cellular immune system.

Dane particles   Virus-like particles found in the serum of hepatitis patients that possess the HBsAg antigen. These particles are composed of a nucleocapsid (28 nm) surrounded by an outer envelope (42 nm).

Decapsidation   The process by which an animal or human virus's nucleic acid is released from its capsid once the virion is inside its host cell. Also called uncoating.

Defective Interfering (DI) Particles   Virus particles whose nucleic acids lack part of the viral genome; DI particles often interfere with the replication of the standard virus.

Dementia   Chronic intellectual impairment (loss of mental capacity) with organic origins, that affects a person's ability to function in a social or occupational setting. See "AIDS Dementia Complex".

Dendritic cell   A type of antigen-presenting immune cell. Dendritic cells have elongated, tentacle like branches in which they trap foreign objects.

Dermal   Relating to the skin.

Diarrhoea   Abnormally frequent and liquid stools.

Didanosine (ddI)   A nucleoside analogue reverse transcriptase inhibitor. Didanosine comes under the same family of antiviral drugs as AZT. Its primary toxicities are pancreatitis and neuropathy.

Dihydrofolate Reductase (DHR) The bacterial (or protozoan) enzyme targeted by trimethoprim-sulphamethoxazole. DHR is necessary for pneumocystis or toxoplasmosis to survive.

Echoviruses   A group of enteroviruses belonging to the Picornaviridae family. The name is an acronym for Enteric Cytopathic Human Orphan viruses.

Eclipse   The time interval following bacteriophage infection when no complete virions are present in the host cell.

Edema   Swelling.

Electron Microscopy   An imaging method, which uses a focused beam of electrons to enlarge the image of an object on a screen or photographic plate.

ELISA (Enzyme Linked Immunosorbent Assay)   A laboratory test to determine the presence of antigens or antibodies by assaying the amount of bound enzyme linked to a specific immunoglobulin.

Enations　Small outgrowths (often on the underside of leaf veins) caused by certain plant viruses.

Encephalitis　A general term denoting inflammation of the brain.

Encephalopathy　Lesions in the brain, or general degeneration of brain matter.

Endemic　Pertaining to diseases associated with particular locales or population groups.

Endocytosis　The process by which cells form a membrane vesicle around a substance and take the material into the cell. This is the general term for phagocytosis and pinocytosis.

Enteric　Relating to, or of the intestines or gastrointestinal tract.

Envelope　Membrane surrounding a virus particle and usually containing virus-encoded proteins.

Enzootic　A disease that occurs or persists primarily in animals.

Enzyme　A protein that functions as a catalyst to speed up the rate of a chemical reaction.

Epidemiology　The science concerned with the determination of the specific causes of a disease or the interrelation between various factors determining a disease, as well as disease trends in a specific population.

Episome　See plasmid.

Epitope   Small groupings of amino acids that elicit the formation of specific antibodies.

Epstein–Barr Virus (EBV)   A herpesvirus that causes infectious mononucleosis and hairy leukoplakia. EBV also has been associated with Burkitt's lymphoma, a cancer of the lymph nodes and the nasopharynx, the part of the pharynx (throat) which lies above the soft palate.

Erythema   Redness of the skin due to capillary dilation.

Erythematous   Red or reddened.

Etiology   The study or theory of factors which cause disease.

Excision   Release of an integrated provirus; reversal of integration. Also removal of mismatched or damaged nucleotides from DNA molecule.

Exponential Phase   The growth phase of an organism during which the maximum growth rate (minimum generation time) for a given set of growth conditions is maintained.

Famciclovir (Famvir)   A drug chemically related to acyclovir, used in the treatment of herpetic diseases.

Febrile   With a fever.

Filovirus   The thread-like virus family which includes such viruses as Ebola and Marburg that are very deadly.

Fluorescent antibody   Antibody conjugated with a fluorescent dye and used in combination with a fluorescence microscope to detect the presence of viral antigens in cells.

Foscarnet (Foscavir)   An antiviral FDA-approved drug for the treatment of cytomegalovirus (CMV) retinitis and other diseases caused by CMV. It is also used to treat acyclovir-resistant herpes virus infections. Adverse side effects may include kidney toxicity, muscle twitching, nausea and skin ulcers.

Fulminating   Serious acute and active infection.

Fusion   Process by which enveloped viruses become incorporated into the cell membrane and thus enter the cell. It is also a mechanism by which such viruses bring cell membranes together to form multinucleated cells or syncytia.

Ganciclovir (Cytovene)   An antiviral FDA-approved drug to treat cytomegalovirus (CMV) retinitis and to prevent CMV disease in transplant patients at risk for CMV. It has also been used to treat CMV colitis, CMV oesophagitis, AIDS-related meningoencephalitis and AIDS-related polyradiculopathy; generally administered intravenously. An oral form of the drug is under study for the treatment and prevention of CMV disease in people with HIV infection.

Gastric   Relating to the stomach.

Gastroenteritis   Inflammation of the mucous membrane of the stomach and/or the small intestine.

Gene   The sequence of nucleotides in DNA that are transcribed into RNA and/or translated into protein.

Generalized transduction   The transfer of randomly selected small pieces of host-cell DNA by defective bacteriophages.

Generation time   The interval between divisions of a growing microorganism or the interval needed by a culture to double its cell number.

Genome The DNA code that comprises the complete genetic composition of an organism.

Gingivostomatitis The cold sores or fever blisters of the lips and mouth resulting from infection by herpes simplex virus type 1 (HSV-1).

Glycoprotein A protein molecule coated with sugars.

Glycoprotein Polypeptide to which at least one carbohydrate residue is covalently attached.

gp120 Another protein on the outer shell, or envelope, of HIV. gp120 is the portion of HIV that binds to a helper T-cell's surface protein, CD4. The "120" refers to its molecular weight.

gp160 The "precursor" protein to both gp41 and gp120, gp160 is cleaved by viral enzymes into the two surface proteins at a late stage of viral assembly.

gp41 A protein on the outer shell, or envelope, of HIV. gp41 is the portion of HIV that pierces a helper T-cell's surface protein, CD4, allowing viral entry. The "41" refers to its molecular weight.

Granulocytes A white blood cell type of the immune system filled with granules of toxic chemicals that enable them to digest microorganisms. Basophils, neutrophils, eosinophils and PMNs are examples of granulocytes.

Granulocytopenia A lack or low level of granulocytes in the blood. Often used interchangeably with "neutropenia."

Haemadsorption Binding of erythrocytes to the surface of virus-infected cells due to the presence of virus-encoded gene products in the cell membrane.

Haemagglutination   The clumping of red blood cells caused by an antibody or a virus.

Haematoma   Bruise.

Haematopoietic   Pertaining to the formation of blood cells.

Haemolysis   Destruction of blood cells.

Haemorrhagic Fever   A condition characterized by non-stop internal or external bleeding resulting from a viral infection which has caused blood vessel damage.

Hairy leukoplakia   A whitish, slightly raised lesion that appears on the side of the tongue. Thought to be related to Epstein–Barr virus infection, it was not observed before the HIV epidemic.

Half life   The amount of time required for half of a given substance to be eliminated from the body.

HeLa cells   Established line of human cervical carcinoma cells often used in studies of the biochemistry and growth of cultured human cells and susceptible to infection by many viruses.

Helix   Spiral structure with a repeating pattern. It is the natural conformation of many regular biological polymers (e.g. polypeptides and polynucleotides) and virus particles.

Hepatic   Pertaining to the liver.

Hepatitis B   A viral liver disease that can be acute or chronic and even life-threatening, particularly in people with poor immune resistance. Like HIV, the hepatitis B virus can be transmitted by sexual contact, contaminated needles or contaminated blood or blood products. Unlike HIV, it is also transmissible through close casual contact.

Hepatitis An inflammation of the liver caused by any of several causes. Often accompanied by jaundice, enlarged liver, fever, fatigue and nausea, and abnormal liver function blood tests.

Hepatitis C A recently recognized viral disease that causes inflammation of the liver, and may cause severe, life-threatening liver damage. Hepatitis C was formerly called "non-A/non-B" hepatitis.

Hepatomegaly An enlargement of the liver.

Hepatosplenomegaly Enlargement of the liver and spleen.

Herpangina A disease caused by coxsackie viruses and marked by vesiculopapular lesion ~1-2 mm diameter, which are present around the fauces and soon break down to form grayish-yellow ulcers.

Herpes labialis A human disease characterized by lesions on the lips caused by HSV-1.

Herpes Simplex Virus 1 (HSV-1) A virus that can cause painful "cold sores" or blisters on the lips ("fever blisters") or in the mouth or around the eyes. The symptomatic disease stage occurs at unpredictable intervals of weeks, months or years. The latent (inactive) virus can reactivate due to emotional stress, physical trauma, other infections, or suppression of the immune system. HSV-1 responds well to treatment with acyclovir.

Herpes Varicella Zoster Virus (HVZ/VZV)   The varicella virus causes chickenpox in children and may reappear in adulthood as herpes zoster. Herpes zoster, also called shingles, consists of very painful blisters on the skin that follow nerve pathways.

Herpes virus   A family of viruses including Herpes simplex I and II, herpes zoster, Epstein–Barr virus, Cytomegalovirus, and the newly discovered Kaposi's sarcoma-associated herpes virus.

Hexamer   Group of six protein subunits on the triangular faces of an icosahedral virus capsid.

HIV-1   The probable cause of AIDS. A human retrovirus of the lentivirus family, notable for long duration of asymptomatic infection, often followed by progressive deterioration of cell-mediated immune function, and eventual opportunistic infection and neoplasm. The median time for an HIV-infected individual to develop full clinical AIDS is probably over ten years (formerly called HTLV III or LAV).

Host   A cell or organism that supports the growth of a parasite or virus.

Hot virus   A virus that has the potential to spread rapidly and therefore must be handled with extreme care. Examples include the ebola virus and hantavirus.

HSV-2   A virus closely related to HSV-1 that causes similar lesions. However, HSV-2 is usually transmitted sexually, and its lesions generally are in the anogenital area.

Human Immunodeficiency Virus or HIV Infection   Infection with the retrovirus that causes the acquired immunodeficiency syndrome (AIDS).

Human Papillomavirus (HPV)   A member of the papova family of viruses. HPV causes warts or nipple-like protrusions. HPV has also been associated with cervical cancer in women and anal cancer.

Humoral Immune Response   The immune response that is mediated by B-cells and involves production of antibodies. Humoral immunity (also known as TH-1 immune response) is associated with the production of the cytokines interleukin-4 and interleukin-10.

Humoral Immunity   The branch of the immune system that relies primarily upon antibodies. See also "Cellular Immunity."

Hydrophobia   Fear of water.

Hypergammaglobulinemia   Abnormally high levels of antibodies in the blood. Common in persons with HIV.

Icosahedron   A symmetrical structure has 12 vertices with 5-fold rotational symmetry; the centre of each triangular face is on a 3-fold symmetry axis, and the midpoint of each edge is on a 2-fold symmetry axis. There are 20 equilateral triangles that enclose a central space.

Icterus   Jaundice caused by liver disease.

Immune deficiency   A breakdown or inability of certain parts of the immune system to function, thus making a person more susceptible to certain diseases to which the person would not ordinarily be subjected. In diseases associated with HIV, cell-mediated immunity related to the function of T-helper lymphocytes deteriorates, increasing the likelihood of disease from a number of pathogens, many of which are ubiquitous (opportunistic infection).

Immune Response   The activity of the immune system against foreign substances.

Immune System   The complex functions of the body that recognize foreign agents or substances, neutralize them, and recall the response later when confronted with the same challenge.

Immunocompetent   Having a normally functioning immune system.

Immunocompromised   Having a deficient or damaged immune response.

Immunoglobulin   A protein that acts as an antibody to help the body fight off disease. There are 5 classes: IgG, IgA, IgD, IgM and IgE. Recombinant and pooled immunoglobulins from blood donations have been used successfully to help HIV-infected children and some adults resist bacterial infections.

In vitro   Latin for "in glass": An artificial environment created outside a living organism, e.g. a test tube or culture plate, used in experimental research to study a disease or process.

In vivo   Latin for "in life". Studies conducted within a living organism, for example animal or human studies.

Inactivated vaccine   Dead microorganisms used as antigens to produce immunity.

Inclusion bodies   Unusual structures occasionally found inside a host cell during virus replication. There are microscopically distinct sites of virus synthesis and/or accumulation of viral products in virus infected cells—often intracellular virus crystals.

Incubation period   The time interval between the initial infection and appearance of the first clinical symptom or sign of disease.

Infection   Condition in which virulent organisms are able to multiply within the body and cause a response from the host's immune defences. Infection may or may not lead to clinical disease.

Infectious mononucleosis   A disease caused by Epstein-Barr virus and characterized by enlarged lymph nodes and many atypical lymphocytes (T cells) in the peripheral blood.

Inflammation   The response to damaged tissue which results in redness, swelling, pain, heat and tenderness.

Injection Drug User (IDU, IVDU, IVDA)   Also known as intravenous drug user (IVDU), or intravenous drug abuser (IVDA). None of these terms are very precise, often IVDU will be used to describe someone who injects IM or Sub-Q, and the distinction between "user" and "abuser" is a controversial and emotionally charged one. Nevertheless, a working definition for an IDU is anyone who regularly injects any substances, whether pharmaceutically or illicitly made, not under medical prescription and supervision.

Integrase   The HIV enzyme that governs the insertion of HIV's proviral genetic material into the host genome. Integrase is a target for a new generation of HIV drugs.

Interferons   A family of secreted proteins (lymphokines) in the body with the ability to induce an antiviral state in most cell types. They are secreted by infected host cells to protect uninfected cells from viral infections. There are 3 main classes of interferons: alpha, beta and gamma. The interferons have been synthesized by genetic engineering, and are being tested as treatments for HIV infections and other diseases. Alpha interferon is FDA-approved for treatment of HIV–related Kaposi's sarcoma, chronic hepatitis B and genital warts.

Interferons   Animal cell glycoproteins, produced by most vertebrate cells that protect other cells from viral infection.

Interleukin   A chemical hormone messenger (cytokine) secreted by and affecting many different cells in the immune system.

Intradermal   Within the layers of the skin.

Intramuscular   Into the muscle, frequently in reference to injections.

Intravenous (IV)   Within or into the veins. Intravenous drugs are injected directly into the veins.

Jaundice   Yellowish discolouration of the skin and eyes due to bile. Usually associated with some form of liver damage or malfunction.

Kaposi's Sarcoma (KS)   A tumour of the wall of blood vessels, or the lymphatic system. Usually appears as pink to purple, painless spots on the skin but may also occur internally in addition to or independent of lesions.

Keratitis   Infection of the cornea; can lead to blindness.

Koplik Spots   Blue-gray specks highlighted by an inflamed background that appear on the inside of the cheek before the measles rash develops.

Latency   The period when an organism is in the body, shows no symptoms, but is in an inactive state (also known as incubation period).

Latent Infection   Viral infection in which the virus responsible is able to avoid the host's immune system and defences.

Latent period   Portion of time in the virus infection cycle between the apparent disappearance of the infecting virus and the appearance of newly synthesized progeny (see Eclipse).

Long Terminal Repeat (LTR)   1) Genetic material at the beginning of the HIV genome that helps control replication. 2) Sequences at the ends of the retroviral genome containing promoters and enhancers necessary for efficient virus replication.

Lymph nodes   Small bean-sized organs of the immune system, distributed widely throughout the body. Lymph fluid is filtered through the lymph nodes in which all types of lymphocytes take up temporary residence. Antigens which enter the body find their way into lymph or blood and are filtered out by lymph nodes or the spleen respectively, for attack by the immune system.

Lymphadenopathy   Swollen, firm and possibly tender lymph nodes. The cause may range from an infection such as HIV, the flu, mononucleosis, or lymphoma (cancer of the lymph nodes).

Lymphocyte   White blood cells that mature and reside in the lymphoid organs, and are responsible for the acquired immune response.

Lymphoma   Cancers of the lymphatic system, often the T- or B-lymphocytes. There are many categories of lymphoma, including lymphoblastic, cleaved, non-cleaved, Burkitt's and Hodgkin's disease. Many lymphomas count as an AIDS diagnosis.

Lyse   To rupture or destroy a cell.

Lysis   The process of lysing or destroying cells.

Lysogen   A bacterium that carries a prophage in its genome.

Lysogenic viruses   Viruses (generally bacteriophages) that can become stably established within the host, either by integration into the host's genome or by plasmid formation, and thereby establish immunity in the host cell to reinfection by the same virus.

Lytic cycle   The viral replication that results in the lysis of the host cell and the release of virulent virions.

Maculopapular Skin Rash   Spotty blemishes with raised inflamed centres.

Malaise   A generalized nonspecific feeling of discomfort and/or fatigue.

Malignant   Cells or tumours growing in an uncontrolled fashion.

Meninges   Membranes surrounding the brain or spinal cord. Part of the so-called "blood–brain barrier".

Meningitis   An inflammation of the meninges.

Metastases   Secondary cancers, usually due to malignant cells carried from the primary cancer in the bloodstream or lymphatic system to other tissues.

Microtubules   Hollow proteinaceous filaments that are structural components of centrioles, eukaryotic flagella, cilia, and the architectural framework of the eukaryotic cytoplasm.

Molluscum contagiosum   An infectious skin condition characterized by small whitish papules, generally on the face or the trunk.

Morbidity   The condition of being diseased or sick.

Mortality   The condition of being dead.

Mosaic   Common symptom of plant virus infections in which the leaves of infected plants show a pattern of dark green and light areas. In the leaves of monocots, these symptoms appear as stripes.

Multiple sclerosis   A progressive degenerative disease of the CNS characterized by the loss of certain brain and spinal cord functions.

Multiplicity of infection (m.o.i.)   Number of infectious virions per cell used to initiate infection.

Myeloma   B cell tumour of the immune system.

Nasopharyngeal carcinoma   Tumor involving the epithelial cells of the nasopharynx and linked to Epstein-Barr virus infection.

Natal   Relating to birth.

Negri Bodies   An aggregation of rabies virus nucleocapsids in the brain tissue of a rabid animal. Usually concentrated in the hippocampal pyramidal cells and less frequently seen in cortical neurons and cerebellar Purkinje cells. Observed as cytoplasmic inclusions with round or oval shape, 1 and 7 mm across, eosinophilic bodies.

Neuraminic acid Nine-carbon sugar derivative that is part of the cellular receptor recognized by orthomyxovirus haemagglutinin.

Neuraminidase An enzyme localized in the surface of certain enveloped virions that cleaves the bond joining sialic acid to cell receptor molecules.

Neutralizing antibody An antibody that neutralizes (renders harmless) the infectivity.

Non Nucleoside Reverse Transcriptase Inhibitor (NNRTI) A class of drugs including nevaripine and delaviridine that targets the same reverse transcriptase enzyme of HIV as the nucleoside drugs like AZT, ddI, etc. NNRTIs have been characterized by strong antiviral activity, followed by extremely rapid acquisition of resistance by exposed virus.

Non-persistent transmission Relationship between certain viruses (e.g. potyviruses) and their insect vectors that is characterized by a short acquisition period and a relatively short period during which the vector is able to transmit the virus. Infection involves virus particles associated with mouthparts and not those that have passed into the gut.

Non-structural protein Virus-encoded proteins that are not part of the virus particle, usually functional during replication.

Nosocomial A term for an infection or illness that was acquired in a hospital, or other health care facility.

Nucleic Acid An organic compound made up of a phosphoric acid, a carbohydrate and a base of purine or pyrimidine; formed in helical chains.

Nucleocapsid   The protein–nucleic acid complex of a virus. Some nucleocapsids are surrounded by a lipid-containing viral envelope.

Nucleoprotein   Complex containing nucleic acid and protein.

Nucleoside   A precursor to the cellular building blocks, nucleotides, from which new DNA is constructed. Nucleosides are phosphorylated to produce nucleotides.

Nucleoside Analogue Reverse Transcriptase Inhibitors   A family of antiviral compounds including AZT, ddI, ddC, d4T, 3TC, FLT, PMEA, and others. These compounds are phosphorylated into nucleotide analogues, and interfere with the activity of the viral enzyme reverse transcriptase.

Oncogene   Cellular or virus-encoded gene whose products are able to transform eukaryotic cells so that they begin to grow like tumour cells.

p24 Antigen   A core protein making up the nucleocapsid of the HIV Virus. p24 was thought at one time to be a surrogate marker for disease progression. Now it is recognized that some long-term asymptomatics have relatively high elevations of p24, while others die never having been positive (anything less than 10 picograms/mole is effectively negative).

Palindrome   DNA sequence that is the same when one strand is read left to right and the other is read right to left (i.e., sequence consisting of adjacent inverted repeats).

ACTAGT
TGATCA

Pandemic   Referring to an epidemic disease of widespread prevalence.

Papilloma   A small neoplasm on the skin. Usually called a wart.

Papillomavirus   A family of papova viruses associated with sexually transmitted diseases, including condylomata. Certain papillomavirus variants have also been associated with cervical cancer, particularly in HIV-infected women.

Papule   A small raised bump or protrusion on the skin.

Parenteral   Not through the mouth. Intravenous, intramuscular, and intradermal administration are all parenteral.

Paresthesia   Abnormal sensations, numbness, tingling, burning.

Pathogen   An organism or virus that causes a disease.

Pathogenesis   The natural evolution of a disease process in the body without intervention (i.e., without treatment); Description of the development of a particular disease, especially the events, reactions and mechanisms involved at the cellular level.

Peplomers   Protrusions from the surface of enveloped viruses (also called spikes).

Perinatal   Relating to the period around the time of birth.

Permissive cells   Cells that support complete virus replication. This term is used in contrast to non-permissive cells that show abortive infection.

Peyer's Patch   Lymphatic tissue lining the intestines. One of the largest areas of lymphatic tissue in the body. May take over some of the role of the thymus in adults.

Phage Conversion   Alterations of a host cell's phenotype resulting from the expression of one or more prophage genes.

Phase I Trial   The first stage in testing a new drug in humans. Performed as part of an approved Investigational New Drug Application under Food and Drug Administration guidelines. The studies are usually done to generate preliminary information on the chemical action and safety of the drug using normal healthy volunteers. Usually done without a comparison group.

Phase II Trial   The second stage in testing a new drug in humans. Performed as part of an approved Investigational New Drug Application under Food and Drug Administration guidelines. Generally carried out on patients with the disease or condition of interest. The main purpose is to evaluate activity, and possibly provide preliminary information on treatment efficacy and to supplement information on safety obtained from phase I trials. Usually, but not always, designed to include a control treatment and random allocation of patients to treatment.

Phase II/III   A special classification arising from the AIDS context where greater testing of efficacy occurs earlier.

Phase III Trial   The third and usually final stage in testing a new drug in humans. Performed as part of an approved Investigational New Drug Application under Food and Drug Administration guidelines. Concerned primarily with assessment of dosage effects and efficacy and safety. Usually designed to include a control treatment and random allocation to treatment. Once this phase is completed the drug manufacturers may request permission to market the drug by submission of a New Drug Application to the Food and Drug Administration, assuming the results of the phase I, II and III trials are consistent with such a request.

Phase IV Trial Generally, a randomized controlled trial that is designed to evaluate the long-term safety and efficacy of a drug for a given indication and that is done with Food and Drug Administration approval. Usually carried out after licensure of the drug for that indication.

Phosphoproteins Proteins containing phosphate groups, most commonly attached to serine, threonine, or tyrosine residues.

Pitch Repeated periodicity in a helical structure.

Plaque The discrete clear region in the lawn or monolayer of cells in which some of the cells have been lysed following the intracellular reproduction of virus.

Plasmids Extrachromosomal elements (usually covalently closed circular DNAs) that replicate autonomously and range in size from $\leq$ 1 kbp to $\geq$ 300 kbp.

Plasmodesmata Cytoplasmic connections through the cell walls of higher plants, analogous to the gap junctions between certain types of animal cells.

Pneumocystis Carinii Pneumonia (PCP) An opportunistic pneumonitis often seen in HIV-infected patients. PCP generally produces a dry, hacking cough. Although previously thought to be a protozoa, and responsive to anti-protozoan treatment, recent genetic analysis suggests that p carinii is closer to the fungi.

Polyprotein   Large primary translation product from which individual functional proteins are subsequently released by proteolytic cleavage.

Post-Partum   After giving birth.

Prenatal   Relating to the period before birth.

Prion   Term applied to the agents responsible for certain neurological diseases of vertebrates (e.g. scrapie). Derived from "proteinaceous infectious particle" and refers to the apparent absence of a genomic nucleic acid.

Prodromal   Pertaining to symptoms indicating the onset of a disease. May include symptoms prior to those adequate for accurate diagnosis.

Progressive Multifocal Leukoencephalopathy   A rapidly degenerative neurological condition associated with HIV, characterized by diffuse gray-matter pallor on CT, and no focal lesions. Thought to be associated with JC Papovavirus.

Prophage   The genome of a temperate phage when it is integrated into its host cell's DNA (e.g. Lambda phage).

Prophylaxis   Treatment intended to preserve health and prevent the occurrence or recurrence of a disease.

Provirus   A virus genome that has been integrated into either the host genome or into a plasmid. Proviral DNA is passively replicated by the host cell machinery and thus transmitted from one cell generation to another.

Reactivation   Activation of a virus, (e.g. herpesvirus) from a latent stage.

Reticulocyte Immature RBCs that are able to synthesize haemoglobin. Cell-free rabbit reticulocyte extracts are often used to study protein synthesis in vitro.

Retina The back part of the eye that senses light and forms images.

Retinitis Inflammation of the retina, linked in AIDS to CMV infection. Untreated, it can lead to blindness.

Retrovirus A class of viruses, which copy genetic material using RNA as a template for making DNA (HIV is a retrovirus.).

Reverse transcriptase Enzyme encoded by retroviruses (as well as hepadna- and caulimoviruses) that is able to synthesize a complementary DNA copy of an ssRNA template and then to convert this DNA molecule to the dsDNA form. It is an essential step in the life cycle of HIV. AZT, ddI and ddC act against reverse transcriptase.

Rise Period The time interval between the release of the first and the last bacteriophage from infected bacteria in a one-step growth experiment.

Rolling circle Mechanism for nucleic acid replication in which the template is a circular molecule (either DNA or RNA). The newly synthesized strand may either be released as a single copy or, more commonly, continue on to form concatemeric progeny molecules.

Sarcoma A malignant tumour of the skin or soft tissues.

Satellite viruses Defective viruses that are replicated only in the presence of a specific helper virus.

Semipersistent transmission   Relationship between a plant virus and its vector characterized by short acquisition period and no latent period, but the vector remains able to transmit the virus for hours to days.

Serologic Test   Any of a number of tests that are performed on the clear, liquid portion of blood (serum). Often refers to a test which determines the presence of antibodies to antigens such as viruses.

Shingles (Herpes Zoster)   A skin condition characterized by painful blisters in a linear distribution on one side of the body that generally dry and scab, leaving minor scarring. Shingles is caused by the re-activation of a previous infection with the varicella-zoster virus that causes chickenpox, usually early in life. Shingles may be a symptom of HIV disease progression. Shingles may recur in people with poor immunity.

Specialized Transduction   The bacteriophage-mediated transfer of host cell genes that are located on the chromosome adjacent to the integration site of a prophage.

Subcutaneous   Beneath or introduced beneath the skin (e.g. subcutaneous injections).

Subgroup   A subpart of the study population distinguished by a particular characteristic or set of characteristics (e.g. males under age 45 at entry).

Subtype   A genetic variant.

Superinfection   Attempt to infect a host with a second virus. May result in interference, synergism, recombination, or phenotype mixing.

Syncytia A mass of cells which fuse together to form one "giant cell." In HIV infection this condition leads to direct cell-to-cell infection and continued HIV replication.

Syndrome A group of symptoms and diseases that together are characteristic of a specific condition.

T antigens Proteins associated with tumour production and cell transformation by various viruses

T Lymphocyte Cells Thymus-dependent cells that coordinate the cell-mediated immune system.

Temperate Bacteriophage A bacterial virus that exists in both a replicative form capable of causing cell lysis and in a genetic form (prophage) that is passed on to the host's daughter cells.

Titer ("Titre") A laboratory measurement of the amount (or concentration) of a given component in solution.

Transduction The virus-mediated genetic transfer of genes from a host cell to a recipient cell.

Transfection Initiation of a virus infection by inoculation with viral nucleic acid; a process similar to transformation.

Transformation Gene transfer via DNA uptake by "competent cells". Also used to describe the morphological changes in cells, often associated with malignancy.

Uncoating Removal of the outer layer(s) of a virus particle following infection and leading to the release of the viral nucleic acid.

Vaccination   The act of administering a vaccine to lessen or prevent the effects of an infectious agent.

Vaccine   A substance that contains antigenic components from an infectious organism. By stimulating an immune response (but not disease), it leads to immunity to a certain microorganism and protects against subsequent infection by that organism.

Varicella-Zoster Virus (VZV)   A virus in the herpes family that causes chickenpox during childhood and may reactivate later in life to cause herpes zoster (shingles) in immunosuppressed individuals.

Vertical Transmission   Transmission of HIV from mother to foetus.

Viral Burden/Load   The concentration of a virus in the body.

Viraemia   The presence of virus in the blood.

Virion   A virus particle existing freely outside a host cell.

Viroid   A very exotic type of virus-like particle that infects only plant cells and consists of a group of membraneless circular RNAs that neither code for nor contain any structural protein. They replicate without other viruses being present.

Viropexis   Putative special form of a phagocytosis by which some viruses enter animal cells through the plasma membrane.

Viroplasm   amorphous cytoplasmic inclusion body associated with certain virus infections; usually not surrounded by a membrane.

Virus   A group of infectious agents characterized by their inability to reproduce outside of a living host cell. Viruses may subvert the host cell's normal functions, causing the cell to behave in a manner determined by the virus.

Wasting Syndrome   A condition among HIV-infected individuals characterized by involuntary weight loss of more than 10% of baseline body weight. Other symptoms may include chronic diarrhoea or chronic weakness and fever for more than 30 days; a CDC AIDS-defining condition.

Western Blot   A laboratory test of blood for specific antibodies; more accurate than the ELISA test, the Western blot is used as a confirmatory test if an HIV ELISA test is positive.

Wild-Type Virus   The customary type of a virus before genetic manipulation or mutation; virus isolated from an individual, as opposed to that from a lab culture.

REFERENCES

Ananthan S, Saravanan P. "Analysis of human rotavirus serotypes in children with acute diarrhea in Chennai by monoclonal antibody based ELISA." *Indian J Med Res.* 1998; 108: 58–61.

Ananthan S, Saravanan P. "Genomic diversity of group A rotavirus RNA from children with acute diarrhea in Chennai, South India." *Indian J Med Res.* 2000; 111: 50–56.

Ballal M, Shivananda PG. "Rotavirus and enteric pathogens in infantile diarrhea in Manipal, South India." *Indian J Pediatr.* 2002; 69: 393–396.

Bishop RF, Masendycz PJ, Bugg H, Carlin JB, Barnes GL. "Epidemiological patterns of rotaviruses causing severe gastroenteritis in young children throughout Australia from 1993 to 1996." *J. Clin. Microbiol.* 2001; 39: 1085–1091.

Bollinger RC, Brookmeyer RS, Mehendale SM, *et al.* "Risk factors and clinical presentation of acute primary HIV infection in India." *JAMA.* 1997; 278: 2085–2089.

Broor S, Husain M, Chatterjee B, Chakraborty A, Seth P. "Direct detection and characterization of rotavirus into subgroups by dot blot hybridization and correlation with 'long' and 'short' electropherotypes." *Clin Diagn Virol.* 1995; 3: 29–38.

Broor S, Husain M, Chatterjee B, Chakraborty A, Seth P. "Temporal variation in the distribution of rotavirus electropherotypes in Delhi, India." *J Diarrhoeal Dis Res.*1993; 11: 14–18.

Broor S. "Recent avian influenza outbreaks: A pandemic in the waiting." *Indian J Med Microbiol.* 2005; 23(2): 72–73.

Brown DWG Mathan MM, Mathew M, Martin R, Beards GM, Mathan VI. "Rotavirus epidemiology in Vellore, South India: group, subgroup, serotype and electropherotype." *J Clin Microbiol.* 1988; 26: 2410–2414.

Carlos CC, Oishi K, Cinco MT, *et al.* "Comparison of clinical features and hematologic abnormalities between dengue fever and dengue hemorrhagic fever among children in the Philippines" *Am. J Trop Med Hyg.* 2005;73(2):435–440.

Chakravarti A. "Measles control: current trends and recommendations." *Indian J Med Res.*2005; 121: 73–76.

Chandy S, Mitra S, Sathish N. "A pilot study for serological evidence of hantavirus infection in human population in South India." *Indian J Med Res.* 2005; 122: 211–215.

Chatterjee B, Husain M, Kavita, Seth P, Broor S. "Diversity of rotavirus strains infecting pediatric patients in New Delhi, India." *J Trop Pediatr.* 1996; 42: 207–10.

Chatterjee S, Chattopadhyay D, Bhattacharya MK, *et al.* "Serosurveillance for Japanese encephalitis in children in several districts of West Bengal, India." *Acta Paediatr.* 2004 Mar; 93(3):390–3.

Chaturvedi UC, Shrivastava R, Nagar R. "Dengue vaccines: problems and prospects." *Indian J Med Res*. 2005; 121: 639–652.

Collier LH, Timbury MC (Eds.). *Topley and Wilsons Principles of Bacteriology, Virology and Immunity*. 8th edition, vol.4, Edward Arnold, a division of Hodder and Stiughton, London, 1990.

Cunliffe NA, Kilgore PE, Bresce JS, Steele AD, Luo N, Hart CA, Glass RI. "Epidemiology of rotavirus diarrhoea in Africa : a review to assess the need for rotavirus immunization." *Bull WHO*. 1998; 76: 525–537.

Das BK, Gentsch JR, Cicirello HG, Woods PA, Gupta A, Ramachandran M, Kumar R, Bhan MK, Glass RI. "Characterization of rotavirus strains from newborns in New Delhi, India." *J Clin Microbiol*. 1994; 32: 1820–1822.

Das S, Sen A, Uma G, Varghese V, Chaudhuri S, Bhattacharya SK, Krishnan T, Dutta P, Dutta D, Bhattacharya MK, Mitra U, Kobayashi N, Naik TN. "Genomic diversity of group A rotavirus strains infecting humans in Eastern India." *J. Clin microbial*. 2002; 40: 146–149.

Dash PK, Parida MM, Saxena P, et al. "Emergence and continued circulation of dengue–2 (genotype IV) virus strains in northern India." *J Med Virol*. 2004 Oct;74(2):314–22.

Dash PK, Saxena P, Abhyankar A, et al. "Emergence of dengue virus type–3 in northern India." *Southeast Asian J Trop Med Public Health*. 2005;36(2):370–7

De A, Navivadekar R, Mathur M, et al. "Prevalence of rotaviral diarrhoea in hospitalised children." *Indian J Med Microbiol*. 2005; 23(2): 67–70.

Dennehy PH. "Active immunization in the United States: Developments over the past decade." *Clin Microbiol Rev.* 2001; 14: 872–908.

Desrosiers RC. "Prospects for an AIDS vaccine." *Nature Med.* 2004; 10: 221–223.

Excler JL. "AIDS vaccine development: perspectives, challenges and hopes." *Indian J Med Res.* 2005; 121: 568–581.

Gajanana A, Thenmozhi V, Samuel PP, Reuben R. "A community–based study of subclinical flavivirus infections in children in an area of Tamil Nadu, India, where Japanese encephalitis is endemic." *Bull World Health Orgn.* 1995;73(2):237–44.

Halstead SB. "Arboviruses of the pacific and Southeast Asia." In Feigin RD, Cherry JD, eds. *Textbook of Pediatric Infectious Diseases*, 3rd edn. Philadelphia, PA: WB Saunders. 1992; 1468–1475.

Husain M, Dar L, Seth P, Broor S. "Characterization of rotaviruses from children with acute diarrhoea in Delhi." *Indian J Med Microbiol.* 1996; 14: 37–41.

Husain M, Seth P, Broor S. "Detection of group A rotavirus by reverse transcriptase and polymerase chain reaction in faeces from children with acute gastroenteritis." *Arch Virol.* 1995; 140: 1225–1233.

Jain V, Das BK, Bhan MK, Glass RI, Gentsch JR. "The Indian Strain Surveillance Collaborating Laboratories. Great Diversity of group A rotavirus strains and high prevalence of mixed rotavirus infections in India." *J. Clin. Microbiol.* 2001; 39: 3524–29.

Japanese encephalitis: World Organization for Animal Health (OIE), 2003; 1–3.

Kabilan L, Edwin N, Balashankar S, Meikandan D. "Japanese encephalitis among paediatric patients with acute encephalitis syndrome in Tamil Nadu, India." *Trans Roy Soc Trop Med and Hyg.* 2000; 94: 157–158.

Kabilan L, Velayutham T, Sundaram B. "Field– and laboratory–based active dengue surveillance in Chennai, Tamil Nadu, India: observations before and during the 2001 dengue epidemic." *Am J Infect Control.* 2004 Nov;32(7):391–6.

Kandathil AJ, Ramalingam S, Kannangai, *et al.* "Molecular epidemiology of HIV." *Indian J Med Res.* 2005; 121: 333–344.

Kelkar SD, Purohit SG, Simha KV. "Prevalence of rotavirus diarrhoea among hospitalized children in Pune, India." *Indian J Med Res.* 1999; 109: 131–135.

Kelkar SD. "Prevalence of human group A rotavirus serotypes in Pune, India (1990–1993)." *Indian J Med Res.*1997; 106: 508–512.

Khetawat D, Dutta P, Gupta S, Chakrabarti S. "Emergence of rotavirus G4P8 strain among children suffering from watery diarrhea in Calcutta, India." *Intervirol.* 2001; 44: 306–310.

Krishnan T, Burke B, Shen S, Naik TN, Desselberger U. "Molecular epidemiology of human rotaviruses in Manipur: genome analysis of rotaviruses of long electropherotype and subgroup I." *Arch Virol* 1994; 134: 279–292.

Kumarasamy N, Vallbhanani S, Flanigan TP, *et al.* "Clinical profile of HIV in India." *Indian J Med Res.*2005; 121: 377–394.

Kurien T, Thyagarajan SP, Jeyaseelan J. "Community prevalence of hepatitis B infection and modes of transmission in Tamil Nadu, India." *Indian J Med Res.* 2005; 121: 670–675.

Mahal A, Rao B. "HIV/AIDS epidemic in India: An economic perspective." *Indian J Med Res*. 2005; 121: 582–600.

Matthew's Plant Virology. R. Hull (Ed.). 4th edition, Academic Press, 2002.

Monath TP. "Dengue: the risk to developed and developing countries." *Proc Natl Acad Sci USA*. 1994; 91: 2395–2400.

Murray PR, Baron EJ, Pfaller MA, *et al.* (Eds). *Manual of Clinical Microbiology*, 7th edition, ASM Press, 1999 pp. 832–1158.

Narayanan M, Aravind MA, Thilothammal N, *et al.* "Dengue fever epidemic in Chennai—a study of clinical profile and outcome." *Indian Pediatr*. 2002; 39: 1027–1033.

Padbidri VS, Wairagkar NS, Joshi GD, *et al.* "A serological survey of arboviral diseases among the human population of the Andaman and Nicobar Islands, India." *Southeast Asian J Trop Med Public Health*. 2002 Dec;33(4):794–800.

Paramasivan R, Mishra AC, Mourya DT. "West Nile virus: the Indian scenario." *Indian J Med Res*. 2003 Sep;118:101–8.

Paranjape RS. "Immunopathogenesis of HIV infection." *Indian J Med Res*.2005; 121: 240 – 255.

Parashar UD, Holman RC, Clarke MJ, Bresee JS, Glass RI. "Hospitalizations associated with rotavirus diarrhoea in the United States, 1993 through 1995, surveillance based on the new TCD–9–CM rotavirus–specific diagnostic code." *J. Infect. Dis*. 1997; 177: 13–17.

Raj P, Bhan MK, Prasad AK, Kumar R, Bhandari N, Jayashree S. "Electrophoretic study of the genome of human rotavirus in rural Indian community." *Indian J Med Res*.1989; 89: 65–68.

Ram S, Khurana S, Khurana SB, Sharma S, Vadehra DV, Broor S. "Bioecological factors and rotavirus diarrhoea." *Indian J Med Res.*1990; 91: 167–170.

Ramachandran M, Das BK, Vij A, Kumar R, Bhambal SS, Kesari N, Rawat H, Bahl L, Thakur S, Woods PA, Glass RI, Bhan MK, Gentsch JR. "Unusual diversity of human rotavirus G and P genotypes in India." *J Clin Microbiol.* 1996; 34: 436–439.

Ramachandran M, Gentsch JR, Parashar UD, Jin S, Woods PA, Holmes JL, Kirkwood CD, Bishop RF, Greenberg HB, Urasawa S, Gerna G, Coulson BS, Taniguchi K, Bresee JS, Glass RI. "The National rotavirus strain surveillance system collaborating laboratories. Detection and Characterization of Novel rotavirus strains in the United States." *J Clin Mirobiol.* 1998; 36: 3223–3229.

Ramamurthy N, Pillai LC, Gunasekaran P. "Influenza activity among the paediatric age group in Chennai." *Indian J Med Res.* 2005; 121: 776–779.

Richman DD, Whitley RJ, Hayden FG (Eds.). *Clinical Virology.* 2nd edition, ASM Press, Washington DC., 2002 p. 357–1240.

Saksena NK, Smit TK. "HAART and the molecular biology of AIDS dementia complex". *Indian J Med Res.*2005; 121: 256–269.

Saravanan P, Ananthan S, Dhamodaran S. Standardisation of an improved ELISA for the detection of rotavirus in faeces and its comparison with a commercial ELISA. *Med Sci Res.*1997; 25: 151–154.

Saravanan P, Ananthan S, Sundaram S. "Comparative analysis of monoclonal antibody-based enzyme immunoassay, modified genome electrophoresis and electron microscopy procedures for rotavirus diagnosis from faecal specimens". *Indian J Med Res.*2001; 113: 78–82.

Shah I, Deshpande GC, Tardeja PN. "Outbreak of dengue in Mumbai and predictive markers for dengue shock syndrome." *J Trop Pediatr.* 2004 Oct;50(5):301–5.

Smith KM (Ed.). *Plant Viruses.* 6th edition, Universal Book Stall, New Delhi, 1992. p. 1–208.

Srey VH, Sadones H, Ong S, *et al.* "Etiology of encephalitis syndrome among hospitalized children and adults in Takeo, Cambodia, 1999–2000." *Am J Trop Med Hyg,* 2002; 66: 200–207.

Thakare JP, Rao TL, Padbidri VS. "Prevalence of West Nile virus infection in India." *Southeast Asian J Trop Med Public Health.* 2002 Dec;33(4):801–5

Tiroumourougane SV, Raghava P, Srinivasan S, Badrinath. "Management parameters affecting the outcome of Japanese encephalitis." *J Trop Pediatr.* 2003; 49(3): 153–156.

Tuksinvaracharn R, Tanayapong P, Pongrattanaman S, *et al.* "Prevalence of dengue virus in aedes mosquitoes during dry season by semi-nested reverse transcriptase–polymerase chain reaction (semi-nested RT-PCR)." *J Med Assoc Thai.* 2004;87 Suppl 2:S129–33

Varshney B, Jagannath MR, Vethanayagam RR, Kothandharaman S, Jagannath HV, Gowda K, Singh DK, Rao CD. "Prevalence of, and antigenic variation in, serotype G10 rotaviruses and detection of serotype G3 strains in diarrheic calves: implications for the origin of G10P11 or P11 type reassortant asymptomatic strains in newborn children in India." *Arch Virol.* 2002; 147: 143–165.

Verma HN (Ed.). *Basics of Plant Virology*. Oxford and IBH Publishing Co. (P) Ltd., New Delhi, 2003. p. 1–219.

Victor TJ, Malathi M, Gurusamy D *et al*. "Dengue fever outbreaks in two villages of Dharmapuri district in Tamil Nadu." *Indian J Med Res*. 2002 Oct;116:133–9.

Vijayakumar TS, Chandy S, Sathish N, *et al*. "Is dengue emerging as a major public health problem?" *Indian J Med Res*. 2005;121(2):100–7.

Willson RA (Ed.). *Viral hepatitis: Diagnosis–treatment–prevention*. Marcel Dekker Inc., New York, 1997 p 1–520.

Zuckerman AJ, Banatvala JE, Pattison JR (Eds.). *Principles and practice of clinical virology*. 3rd edition, John Wiley and Sons, New York, 1994.

INDEX

S

T

U

V